LEVERAGING THEMATIC CIRCUITS FOR BIMSTEC TOURISM DEVELOPMENT

DECEMBER 2022

BIMSTEC

ADB

© 2022 Asian Development Bank
6 ADB Avenue, Mandaluyong City, 1550 Metro Manila, Philippines
Tel +63 2 8632 4444; Fax +63 2 8636 2444
www.adb.org

Some rights reserved. Published in 2022.

ISBN 978-92-9269-915-4 (print); 978-92-9269-916-1 (electronic); 978-92-9269-917-8 (ebook)
Publication Stock No. TCS220549-2
DOI: http://dx.doi.org/10.22617/TCS220549-2

Corrigenda to ADB publications may be found at http://www.adb.org/publications/corrigenda.

Notes:
In this publication, "$" refers to United States dollars, "₹" refers to Indian rupees, "NRs" refers to Nepalese rupees, and "SLRs" refers to Sri Lanka rupees..

ADB recognizes "China" as the People's Republic of China, "Korea" as the Republic of Korea, "Russia" as the Russian Federation, and "Turkey" as Türkiye.

The main work for the preparation of this report was undertaken in 2020 and 2021, while travel and tourism was substantially down among most BIMSTEC countries due to the pandemic and some limited recovery was beginning. Periodic updates were made in 2021 and to the third quarter of 2022 based on inputs from the BIMSTEC member states. However, since updated data was not consistently available, some sections may not reflect the most current status of travel and tourism in the region and among certain BIMSTEC members.

The ADB study team for this report was led by Dongxiang Li with support of Lani Garnace and lead ADB sustainable tourism consultant Scott Wayne.

Cover design by Francis Manio.

Printed on recycled paper

Contents

Appendixes

Tables and Figures

Figures

Foreword

Tourism is an important driver for socioeconomic growth and development. Apart from generating millions of jobs and businesses, tourism contributes significantly to the preservation and promotion of natural and cultural heritage, the restoration and maintenance of a clean and green environment, and the development of vital infrastructure, including road, rail, river, sea, and air connectivity.

Development of the tourism sector can also help foster peace and harmony through enhanced community interaction and a greater understanding and respect for multiculturalism engendered by increased interaction.

The Bay of Bengal Initiative for Multi-Sectoral Technical and Economic Cooperation (BIMSTEC) is made up of Bangladesh, Bhutan, India, Nepal, and Sri Lanka from South Asia; and Myanmar and Thailand from Southeast Asia. Steeped in history, culture, and tradition, the region is home to over 170 UNESCO World Heritage sites. It is renowned for its breathtaking landscapes, majestic mountains, beautiful beaches, lush forests, and rich biodiversity. The region is also home to the largest sea and land mammals: blue whales and elephants. The BIMSTEC region caters to a diverse range of interests and is a treasure trove for tourists, both from within and outside the region.

Recognizing the tremendous potential for cooperation in the tourism sector and the benefits that would accrue to the region, BIMSTEC Leaders had identified tourism as one of the first six areas for regional cooperation. They emphasized the importance of harnessing the region's natural, cultural, and historical highlights, and promoting intra-BIMSTEC tourism, including specialized tourist circuits and eco-tourism. The Second BIMSTEC Tourism Ministers' Roundtable held in Kathmandu, Nepal in 2006 adopted the Plan of Action for Tourism Development and Promotion for the BIMSTEC Region.

The BIMSTEC region experienced continuous growth in tourist numbers and the tourism industry for more than a decade until the outbreak of the COVID-19 pandemic in 2019. The travel restrictions and lockdown measures imposed globally to contain the spread of the virus had devastating consequences for the tourism industry in all the BIMSTEC member states. Although the pandemic has waned, the tourism industry in the entire region is still struggling to recover.

This is an opportune moment to recognize the importance of enhancing regional cooperation for the rebuilding and recovery of the tourism sector. Collaboration among member states should be encouraged and reinvigorated through sharing of information, know-how, best practices, capacity building, and public-private partnership (PPP). Innovative and creative thinking, including the introduction of new policies and exploring novel practices to revive the tourism sector in the region, must be explored.

As such, this report is a pertinent and timely initiative. It analyzes the main tourism drivers and trends, along with the impact of the COVID-19 pandemic. Furthermore, it outlines short-, medium- and long-term recommendations for action. For the robust and resilient recovery of the sector as well as sustainable, inclusive, and integrated development of regional tourism, the BIMSTEC member states should consider implementing the appropriate inputs and suggestions provided in the report.

As member states prepare an updated Plan of Action for Tourism Cooperation in the BIMSTEC region, I know this study report will provide valuable insights.

I would like to express my appreciation to all BIMSTEC member states for their cooperation and contributions to help finalize this report. I also thank the Asian Development Bank and its staff for their support in producing this important publication.

Tenzin Lekphell
Secretary General
BIMSTEC

Preface

The Asian Development Bank (ADB) is pleased to support the Bay of Bengal Initiative for Multi-Sectoral Technical and Economic Cooperation (BIMSTEC) Secretariat in preparing this report on *Leveraging Thematic Circuits for BIMSTEC Tourism Development*. As a regional development bank mandated to promote regional cooperation and integration among its developing member countries, ADB values its partnership with BIMSTEC.

ADB has been cooperating with BIMSTEC since 2005. Through regional technical assistance, ADB supported the preparation of this report to facilitate intraregional cooperation in this important sector with multiple stakeholders from the public and private sectors. ADB has also supported the strengthening of the BIMSTEC Secretariat, which leads member states toward joint actions to promote tourism in the region. Among the joint actions, the members vowed to boost tourism in the region by nurturing closer cooperation in transport networks, creating intraregional thematic tour packages, simplifying cross-border immigration procedures, sharing best practices, and developing an updated action plan.

This report sets the stage for developing a comprehensive tourism strategy and action plan to address issues including the severe impacts of the COVID-19 pandemic on intraregional and inbound tourism in the BIMSTEC region. The report emphasizes further developing thematic circuits as an important product offer for BIMSTEC member states to boost economic recovery from the pandemic. Thematic circuits can be effective ways to organize and manage sustainable tourism since they can bring together public and private sector stakeholders to improve destinations along circuit routes. The BIMSTEC Secretariat could be the central catalyst for realizing this vision, addressing the issues, achieving regional objectives, and maximizing the benefits of thematic circuit–driven tourism for all member states.

I would like to acknowledge the extensive efforts of the report team from the Regional Cooperation and Operations Coordination Division, South Asia Department. I would also like to thank the more than 200 tour operators and officials in BIMSTEC Secretariat and member states who contributed greatly to this report through surveys, interviews, and shared information. The process began just before COVID-19 started to impact the travel and tourism industry, which created an unprecedented set of challenges in researching and writing this report, all of which could not have been possible without the cooperation of these public and private sector stakeholders. Thanks also go to relevant ADB resident missions, the Southeast Asia Department, the Sustainable Development and Climate Change Department, and the Department of Communications for providing substantial comments to improve the report. All their contributions are gratefully acknowledged.

K. Yokoyama

Kenichi Yokoyama
Director General
South Asia Department
Asian Development Bank

Abbreviations

ADB	Asian Development Bank
ASEAN	Association of Southeast Asian Nations
ATTA	Adventure Travel Trade Association
BIMSTEC	Bay of Bengal Initiative for Multi-Sectoral Technical and Economic Cooperation
CBT	Community-based tourism
CDC	Centers for Disease Control and Prevention
EU	European Union
FDI	foreign direct investment
GDP	gross domestic product
GMS	Greater Mekong Subregion
GSTC	Global Sustainable Tourism Council
IATA	International Air Transport Association
KPI	key performance indicator
LTC	Leave Travel Concession
MVA	motor vehicle agreement
NTDP	National Tourism Development Plan
PPP	public–private partnership
PRC	People's Republic of China
SAARC	South Asian Association for Regional Cooperation
SASEC	South Asia Subregional Economic Cooperation
SDG	Sustainable Development Goal
SLTDA	Sri Lanka Tourism Development Authority
SME	Small and medium enterprise
SWOT	strengths, weaknesses, opportunities, and threats
UNDP	United Nations Development Programme
UNESCO	United Nations Educational, Scientific and Cultural Organization
UNWTO	UN World Tourism Organization
USAID	United States Agency for International Development
WEF	World Economic Forum
WEF-TTCI	World Economic Forum Travel and Tourism Competitiveness Index
WTTC	World Travel & Tourism Council

Executive Summary

The countries of the Bay of Bengal Initiative for Multi-Sectoral Technical and Economic Cooperation (BIMSTEC) region are rich in assets and activities that attract visitors from around the world. BIMSTEC leaders agreed on joint actions to promote tourism in the region at ministerial meetings held in February 2005 and August 2006. At the 2005 meeting, they vowed to increase tourism among their countries by fostering closer cooperation in transport networks, creating intraregional thematic tour packages, simplifying cross-border immigration procedures, sharing best practices, and implementing an action plan to realize these objectives. This commitment was reaffirmed at the 2006 meeting, albeit with a revised plan.

This report sets the stage for developing a comprehensive tourism strategy in the BIMSTEC region that includes an updated action. It identifies and proposes ways to address issues and needs in the tourism sector, including the impacts of the coronavirus disease (COVID-19) pandemic, which has hit intraregional and inbound tourism in the BIMSTEC region.

Tourist arrivals and receipts in the BIMSTEC region grew more than 250% since 2010 to almost $66 million and $99 billion, respectively, by 2019. Parallel to this, intraregional (Section 3.1) and domestic tourism growth among BIMSTEC countries had been steady. Factoring in some of the main drivers of global and regional tourism, especially segments such as cultural and wellness tourism, and rising income levels among BIMSTEC countries, the future looked promising for a return to the growth trajectory before the outbreak of COVID-19.

BIMSTEC countries can offer thematic tourist circuits to boost economic recovery from the pandemic and a return to pre-COVID-19 tourism growth. Circuits are essentially itineraries that many tour operators offer throughout the region. They are effective ways to organize and manage sustainable tourism within and between countries since they can bring together public and private sector stakeholders to improve destinations along circuit routes. Fully realizing the benefits of cross-border tourist circuits, however, requires tackling multiple issues throughout the region. BIMSTEC's Secretariat could, with sufficient resources, be the central catalyst for realizing this vision, addressing the issues, achieving regional objectives, and maximizing the benefits of thematic, circuit–driven tourism for all member states.

COVID-19 had begun its rapid global spread by early February 2020. By mid-March, travel and tourism around the world was collapsing, especially in major tourism markets, causing widespread cancellations, job losses, and business failures. With vaccine programs commencing in late 2020 and 2021 to bring COVID-19 under control, the prospects for some recovery in 2022 seemed brighter, with travel and tourism potentially on a faster recovery track. Increased bilateral and multilateral cooperation will be essential to make this happen.

Tourism was growing rapidly in Asia and the Pacific—and, indeed, globally—before the COVID-19 outbreak. In 2019, global tourism arrivals were up 3.1% from 2018 and arrivals to Asia and the Pacific rose 4.1% (371 million arrivals), generating receipts of $1.5 billion globally and $445 million in Asia and the Pacific. Among BIMSTEC countries, arrivals increased from 63.3 million in 2018 to 65.0 million in 2019, generating receipts of $93.2 billion and almost $99.0 billion, respectively. Thailand accounted for the largest share on almost 40 million arrivals and $61.5 billion in receipts. Intraregional travel among most BIMSTEC countries was also growing steadily, marked by substantial numbers of outbound visitors from India to Bangladesh, Bhutan, Myanmar, and Thailand and from Bangladesh to India and Thailand.

Income growth in BIMSTEC countries before the outbreak was driving some intraregional travel and contributing to overall gross domestic product (GDP) and tourism-related GDP growth. Domestic tourism spending was also growing, especially in India, which dominated the region with receipts of $143 billion in 2019, followed by Thailand with ($28.5 billion). This spending set the stage for domestic tourism in 2020 in these countries and the rest of BIMSTEC and has become the main source of tourism spending during the COVID-19 pandemic.

Multiple demand drivers also generated or facilitated this growth, particularly adventure, cultural, culinary, rural, community-based, and wellness tourism, as well as theme parks and attractions, destination hotels, airline growth, cruises, and shopping.

COVID-19 stopped nearly all international tourism to and among BIMSTEC countries from April to December 2020. The effect on employment and businesses has been severe, resulting in millions of job losses and many temporary and permanent business closures. Taking stock of these impacts has been a priority action area for recovery, which BIMSTEC's Secretariat can assist via ongoing assessments.

For BIMSTEC countries to recover and return to tourism's pre-COVID-19 upward trajectory, and particularly for creating thematic circuits, this report focuses on COVID-19 recovery, infrastructure, marketing and products, human resources, policy, and investment areas. The report recommends actions for each.

The top priority is controlling the spread of COVID-19 and advancing the recovery through vaccination programs and implementing health and safety protocols. International tourism and visitor confidence will be restored as COVID-19 recedes. Creating circuits could expedite a return to pre-COVID-19 tourism growth in BIMSTEC countries because it would help advance cooperation on the joint cross-border implementation of health and safety protocols and standards, as well as on regional tourism, marketing, and branding. Recovering lost tourism jobs and businesses is a priority.

Various infrastructure challenges will need to be tackled to tap the region's tourism potential and cross-border thematic circuits, including areas of tourism and transport infrastructure. The Secretariat has an important role in helping member countries with each area of infrastructure in this effort, especially in sharing best practices and harmonizing standards and regulations across borders.

BIMSTEC tour operators are actively promoting circuits, but regional coordination and government support for infrastructure in fully creating these circuits will be important for tourism. Itineraries for these circuits include the Buddhist circuit for Nepal and India, the Himalaya circuit between Nepal and Bhutan, the river cruise circuit between Bangladesh and India, the ocean cruise circuit between Sri Lanka and Kerala in India, and the heritage site circuit from Thailand to Myanmar. These are discussed in Section 4.

The BIMSTEC region lacks joint branding, market research, and marketing at the government level. The lack of online marketing is an especially serious gap that needs to be filled—a function that could be facilitated by the Secretariat. The private sector is an important partner in this process, especially when individual business objectives align with the broader destination objectives of governments.

Human resource needs for tourism include training on COVID-19 measures and expanded tour guide training. Collaboration on this with the private sector will be essential for this training, especially so that guides can work across BIMSTEC borders.

Policy, governance, and investment in tourism encompass multiple challenges that will need to be addressed for tourism in the region to recover and to resume focus on creating circuits so that they can become a major source of sustainable tourism. These issues include easing visas and liberalizing the aviation industry, increasing the frequency of BIMSTEC Tourism Working Group meetings, and strengthening the capacity of the Secretariat. For increasing investments in tourism, the Secretariat could help with aggregating information on regulations, incentives, and opportunities across the region.

Features of regional tourism plans from the Association of Southeast Asian Nations (ASEAN), the Greater Mekong Subregion Economic Cooperation Program (GMS), South Asian Association for Regional Cooperation, and South Asia Subregional Economic Cooperation are considered and, where possible, referenced in the action plan. The ASEAN and GMS plans have potentially useful inputs for the Plan of Action, including strategic plans that focus on digital marketing and enhanced visitor experiences, multicounty thematic tour programs, training for tourism officials, and branding.

Examples from each BIMSTEC member country are also cited as illustrations of possible best practices for replication or adaptation, such as India's Swadesh Darshan thematic circuit program, which provides a useful toolkit for circuit development.

This report proposes an updated vision and a new mission statement to position the BIMSTEC region as a global center for experiential circuit-based tours. The Secretariat could coordinate public and private sector stakeholders, as well as activities, to realize the vision, achieve strategic objectives, and implement the recommended action areas.

These and the strategic objectives for the Plan of Action for Tourism Development and Promotion for the BIMSTEC Region address each issue. The principles of sustainable tourism from the United Nations World Tourism Organization and the Sustainable Development Goals will be foundational for advancing sustainable tourism in BIMSTEC countries. The core action area is establishing a central online repository of best practices and lessons learned to be maintained by the Secretariat with inputs uploaded by tourism authorities in member countries and the private sector. The private sector would be incentivized to do this by marketing, promotion, and online booking functionality that could be built into the repository and a corresponding website or mobile app.

The following are summary lists of the strategic objectives, and short-, medium-, and long-term action areas:

Short-term (2023–2024) Strategic Objectives and Corresponding Action Areas

SO1: Control, mitigate COVID-19 impacts

- Exchange best practices and policies.
- Adopt healthy industry labels, for example, the Safe Travels global protocols of the World Travel & Tourism Council.

SO2: Recover lost tourism jobs and business

- Conduct public–private assessment of losses.
- Facilitate public–private partnerships (PPPs) for domestic marketing.
- Facilitate government assistance—loans, training, temporary jobs, financing, and health and safety measures.

SO3: Improve tourism infrastructure

- Establish a central online repository and app with tourism-related infrastructure projects.

SO4: Stimulate sustainable product development, marketing of circuits

- Establish an inventory of assets for circuits in the online repository and mobile app.
- Exchange domestic marketing campaign strategies.
- Promote a BIMSTEC vision, mission, and brand as a global circuits leader.

SO5: Human resources development

- Increase "tourism consciousness" for communities in developing circuits.
- Establish a portal for online tourism training and capacity building.
- Conduct needs assessments.

SO6: Maximum quality in governance, policy, and investment

- Coordinate tourism governance, increase the frequency of BIMSTEC Tourism Working Group meetings.
- Establish a BIMSTEC Tourism Emergency Response and Recovery Group.

Medium-Term (2024–2025) Strategic Objectives and Corresponding Key Action Areas

SO3: Improve tourism infrastructure

- Facilitate PPPs for financing, developing, and maintaining infrastructure projects.
- Establish roadside amenities complexes.
- Facilitate hotel development along the circuits.
- Adopt and adapt India's Swadesh Darshan program for circuits.
- Coordinate implementation of the Transport Connectivity Master Plan.
- Facilitate coordination tourism and transport stakeholders to boost circuit development.
- Facilitate restart of low-cost carrier operations via route market.
- Facilitate implementation of the BIMSTEC Motor Vehicle Agreement.
- Review the Motor Vehicle Agreement for "Carnet de Passages en Douane customs documents" access.

SO4: Stimulate sustainable product development, marketing of circuits

- Post lists of tour operators from each BIMSTEC country on the online repository.
- Exchange best practices and lessons learned on confidence building and awareness.
- Establish a market research dashboard.
- Create a full tourism strategy for coordinated regional marketing and promotion to reestablish markets.
- Formulate an outreach strategy to connect with film location scouts and production companies.
- Include sample regional circuit tour packages on the online repository.
- Create product strategies for adventure, wellness tourism, cruise tourism circuits, and ecotourism.
- Use the $70,000 BIMSTEC Tourism Fund for joint marketing activities.
- Establish a BIMSTEC business or general travel card.
- Encourage the production and marketing of virtual audio and video tours.
- Conduct familiarization tours for media and tour operators.
- Organize student exchanges among BIMSTEC countries as well as nonmember countries.
- Create a BIMSTEC tourism marketing template focused on circuits.

SO5: Human resources development

- Conduct an online forum on human resource cooperation to exchange curricula and arrange facility visits.
- Expand the Hazard Analysis and Critical Control Points certification for food safety and training for International Organization for Standardization 45001.
- Adopt and adapt GMS short-term training courses for use by BIMSTEC countries.
- Conduct the United Nations Educational, Scientific and Cultural Organization Cultural Heritage Specialist Guides Training and Certification for World Heritage Sites.

SO6: Maximum quality in governance, policy, and investment

- Establish a forward-looking section on the online repository for policies, protocols, and best practices.

Long-Term (2026 and Beyond) Strategic Objectives and Corresponding Key Action Areas

SO3: Improve tourism infrastructure

- Facilitate government support for infrastructure improvements along regional circuits.
- Encourage Open Skies agreements among BIMSTEC members to attract airlines back to the region.

SO4: Stimulate sustainable product development and marketing of circuits

- Expand regional marketing and promotion of BIMSTEC-branded tourism circuits via a regional BIMSTEC tourism strategy.

SO5: Human resources development

- Conduct circuit training via an online circuit development academy.

SO6: Maximum quality in governance, policy, and investment

- Support the full staffing of BIMSTEC with an expanded budget.
- Expand air connectivity via an expanded regional aviation market.
- Promote PPP opportunities for investment and small and medium-sized enterprise on thematic circuits.

Introduction

Background on Bay of Bengal Initiative for Multi-Sectoral Technical and Economic Cooperation and Tourism

The Bay of Bengal Initiative for Multi-Sectoral Technical and Economic Cooperation (BIMSTEC) comprises one of the largest population blocs in the world. With nearly 1.5 billion people, or 21% of the global population, these countries generate a gross domestic product (GDP) of $2.7 trillion and, pre-COVID-19, were averaging annual economic growth rates of 5.5%. With increased economic integration and cooperation among BIMSTEC's economies, growth could soar, improving livelihoods substantially throughout the region. Given the region's great cultural and natural diversity, the tourism potential, particularly for thematic circuits, is substantial—but it is still relatively untapped.

The countries of the BIMSTEC region–Bangladesh, Bhutan, India, Myanmar, Nepal, Sri Lanka, and Thailand–are rich in assets and activities that attract visitors from around the world. BIMSTEC leaders, at the group's first Tourism Ministerial Meeting in Kolkata in 2005, announced the objective of increasing tourism in their countries via closer cooperation in transport networks, creating intraregional thematic tour packages, simplified cross-border visa procedures, shared best practices, and a proposed action plan to realize their objectives. At the Second BIMSTEC Tourism Ministers' Roundtable and Workshop in 2006 in Kathmandu, the ministers agreed to the Plan of Action for Tourism Development and Promotion for the BIMSTEC Region.

For over a decade before the outbreak of COVID-19, tourist arrivals and receipts grew steadily in every BIMSTEC country. Arrivals rose 250% to almost 66 million from 2010 to 2019 and receipts rose 6% to $99 billion in that period (Table 6). In parallel to this was the steady intraregional growth among BIMSTEC countries (Section 3). In addition, based on available data (Table 16), domestic tourism spending increased 33% in India, 70% in Bangladesh, and 75% in Thailand from 2010 to 2019. Add to this, the popularity of cultural tourism, which the United Nations World Tourism Organization (UNWTO) estimates could be more than a third of international trips, the potential is strong for continued growth, particularly of tourism circuits.[1]

Recovering from the COVID-19 pandemic and returning to this growth trajectory requires strong cooperation among BIMSTEC member states since international travel mostly stopped in 2020 and 2021. This report recognizes that the BIMSTEC Secretariat has an important role in bringing members together for regional cooperation in tourism and marketing cross-border thematic circuits. The Secretariat has had minimal resources to do this, but with the challenges faced by members in restarting their tourism sectors, there is clearly a need to increase its capacity to assist members with the recovery, particularly through the establishment of cross-border circuits. Recommendations on updating the Plan of Action for Tourism Development and Promotion for the BIMSTEC Region to meet current and projected needs are proposed in this report.

---

[1]   UNWTO. 2018. *Tourism and Culture Synergies*. Madrid, p. 74.

Intraregional tourism flows between BIMSTEC countries were already substantial pre-COVID-19. The most popular outbound destination for Bangladesh was India (2.2 million arrivals in 2018), likewise for Nepal (200,438 in 2018), while Bangladesh was the most popular destination for India (2.25 million in 2018).[2] The survey of tour operators in BIMSTEC countries done for this report in 2020 indicated substantial interest in thematic circuits, at least on a country basis. Tour operators that responded to the survey were either already offering or were interested in offering circuit-based itineraries in their countries to international visitors. As cross-border travel resumes, governments have a role in facilitating these pairings and connections. This can include measures to improve infrastructure via coordinated joint marketing, support for human resources via sharing education and training resources, investment incentives, and harmonizing COVID-19 measures and protocols. The Secretariat can play an important role in facilitating regional cooperation in these areas through knowledge sharing and transnational policy coordination and by encouraging the creation of a BIMSTEC destination brand and through joint marketing.

As of February 2020, COVID-19 had begun to spread rapidly around the world. By mid-March, travel and tourism had collapsed, especially in major tourism generating markets, causing widespread cancellations, job losses, and business failures. By April, the UNWTO reported that all destinations worldwide had imposed travel restrictions. Over the rest of 2020 and into 2021, international travel recovered on a very limited basis while domestic travel grew significantly in some countries. Every region was still reporting substantial contractions as of May 2021 in year-to-date international arrivals compared with 2019: –84.7% in Europe and –95.5% in Asia and the Pacific, with Southeast Asia –97.8% and South Asia –89.4%.[3] As of August 2021, the Centers for Disease Control and Prevention (CDC) had issued Level 4 warnings against all travel to 74 economies, including BIMSTEC countries Bangladesh, Nepal, and Thailand (Figure 7); Sri Lanka was at a Level 3 (high risk: recommended travel only if vaccinated), India was Level 2 (moderate risk), and Bhutan on Level 1.[4]

As of August 2021, international travel, was still difficult. The Delta COVID-19 variant was causing more outbreaks, although there had been signs of improvement in reducing the spread in BIMSTEC countries—for example, India's COVID-19 was at Level 2 as vaccination levels increased to 32% (as of 19 August 2021). Thailand, however, was at Level 4 in August 2021, despite lower levels earlier in the year (foontnote 4). Despite the unpredictability and volatility of international travel, India and Thailand, among other BIMSTEC countries, have been successfully focusing on domestic markets. Until the international market goes back to normal, domestic tourism provides an optimal stopgap for governments in the region to review and revise their tourism plans and strategies to prepare for what is likely to be an even more competitive travel market post-COVID-19. The Secretariat could assist in knowledge sharing to help members track and coordinate on health and safety protocols, and entry and travel requirements, especially among BIMSTEC members. When COVID-19 recedes, and traveling becomes easier and safer, cross-border thematic circuits will likely be a product and marketing strategy for the region. The Secretariat can facilitate increased regional cooperation to help make this happen.

---

2   UNWTO. 2019 Yearbook of Tourism Statistics dataset [Electronic]. Madrid. (accessed 20 September 2020).
3   UNWTO. World Tourism Barometer. July 2021. p. 3. Madrid.
4   Centers for Disease Control and Prevention. COVID-19 Travel Recommendations by Destination (accessed 20 August 2021).

Scope, Structure, Methodology, and Sources of the Report

This report starts with a summary of pre-COVID-19 tourism trends globally and in the Asia and the Pacific and the BIMSTEC region. It examines the issues hindering growth and offers recommendations for actions to be taken both while COVID-19 is affecting the sector and for the longer-term post-pandemic recovery. Intraregional thematic circuits are considered as a path to economic recovery and increased regional cooperation.

The report provides recommendations for the Secretariat to assist member countries in increasing tourism growth, ideally via regional cooperation and, once it is safe to travel again, via transnational theme-based circuits as a core offering. Following a background section on global, Asia and the Pacific (Section 2), and BIMSTEC tourism trends (Section 3), the report does a more detailed analysis of the tourism market in the BIMSTEC region. This includes a situation analysis of the effects of COVID-19 in the context of four issues: infrastructure, product and marketing, human resources, and governance, policy, and investment.[5]

The 2006 Plan of Action for Tourism Development and Promotion for the BIMSTEC Region (Section 4) has 14 action points summarized in Table 1. The plan is examined further in Section 4.1.

Table 1: Summary of the 2006 Plan of Action for Tourism Development and Promotion for the Bay of Bengal Initiative for Multi-Sectoral Technical and Economic Cooperation Region

Action Title	Proposed Actions	Status
BIMSTEC Information Center	Set up a BIMSTEC Tourism Information Center in India for publicity and collateral materials.	The Second Meeting of the BIMSTEC Network of Tour Operators held virtually 8–9 December 2020 in Colombo recommended that a virtual center be set up in India.
BIMSTEC Tourism Fund	Set up a fund for the information center to undertake tasks based on the action plan. Each member to contribute $10,000.	The fund was established, but not operationalized.
Tour packages	Tour packages (for two or more countries), including the Buddhist circuit, ecotourism, adventure tourism, and meetings, incentives, conferences, and exhibitions tourism, to be finalized by private stakeholders and promoted among member states.	BIMSTEC tour operators organized and offered a variety of multi-country itineraries, many of which were marketed by overseas operators. Extensive recovery of business will be needed.
Organizing FAM trips	Familiarization trips (two for each member country) for journalists and tour operators will be organized by each member states.	Familiarization trips were standard offers pre-COVID-19.
Travel facilitation	Expert meeting on a BIMSTEC Business Travel Card to simplify travel visas and immigration procedures.	This did not happen, but post-COVID-19 variations of this idea are expected and will require regional coordination.
Student exchanges	Member countries to facilitate student exchanges.	This did not happen at a BIMSTEC level. Post-COVID-19, this will depend on restrictions among states.

continued on next page

---

[5] Infrastructure includes tourism infrastructure (hotels, restaurants, cafes, and visitor centers); air and ground transportation; and health and safety measures. Governance and policy include consideration of visa arrangements, aviation policy, investment information, and BIMSTEC institutional arrangements and capacity.

Table 1 continued

Action Title	Proposed Actions	Status
Parity in entrance fees at archaeological sites	Parity on entrance fees for visiting archaeological sites for nationals of BIMSTEC states.	India has initiated this for some archaeological sites.
Extend accessibility by air, land, and water	Member countries to facilitate more access.	This was happening on a bilateral basis between states. COVID-19 put all of this on hold.
Joint investment promotion	Nepal to compile information on investment opportunities and incentives.	Not yet operationalized.
Human resource	Member states to share information on tourism-related training facilities.	Not yet operationalized.
Crisis management	Nepal to work out the operational modalities of the regional network on crisis management.	This has not yet been operationalized, but as most BIMSTEC countries continue to deal with COVID-19, this is a high priority.
Support from development partners	ADB to examine the proposed technical assistance and the Tourism Information Center will follow up with ADB.	ADB has followed up directly with BIMSTEC since the center does not exist yet.
BIMSTEC Tourism Working Group	Establish the group to decide on the implementation of the Plan of Action for Tourism Development and Promotion for the BIMSTEC Region this action plan, including use of the fund.	The working group is expected to become more active. Terms of reference have not yet been drafted.
Bangladesh to host the Third BIMSTEC Tourism Ministers Roundtable	Roundtable and workshop in November 2007.	This did not happen in November 2007 in Bangladesh. Postponed many times.

ADB = Asian Development Bank, BIMSTEC = Bay of Bengal Initiative for Multi-Sectoral Technical and Economic Cooperation.
Sources: BIMSTEC and Asian Development Bank.

This report is based on desk research and communications, mostly virtual sessions, with government officials and industry representatives in BIMSTEC countries. Because of the COVID-19 pandemic, in-person meetings were not conducted, and it should be noted that the crisis slowed the process of obtaining offline official documentation and inputs. Data on visitor numbers and tourism receipts for all destinations along existing and potential circuits were unavailable, thereby highlighting an important area for both governments and the private sector to address. This report is intended to form the basis for moving toward a BIMSTEC regional tourism strategy that enables thematic circuit opportunities.

A survey was also conducted with tour operators from April to August 2020 with stakeholders in all BIMSTEC countries.[6] A survey of government tourism officials in BIMSTEC countries was conducted online as part of a virtual workshop on 25 November 2020. The survey received 10 responses from officials from all seven BIMSTEC countries. Four countries provided single responses, while government tourism organizations in Bhutan, Sri Lanka, and Thailand each had two official respondents. Both surveys included questions about pre-pandemic tourism.

---

[6]   The survey was conducted via Google Documents.

The World Economic Forum's 2019 Travel and Tourism Competitiveness Index (WEF-TTCI) rankings provided useful inputs to this report.[7] The WEF-TTCI is a biennial analysis and benchmarking of tourism competitiveness in 140 countries conducted via an extensive global survey of industry stakeholders. The rankings span four subindexes of enabling environment, conditions enabling travel and tourism policy, infrastructure, and natural and cultural resources, as well as 14 sets of indicators totaling 87 key performance indicators (KPIs).[8] These KPIs are referenced for selected sections, which can help frame and prioritize some of the issues affecting the establishment of circuits. A World Economic Forum representative informed that the 2021 index would be delayed. The 2021 index is expected to account, as best as possible, for COVID-19 impacts, thus probably also producing significant changes in the rankings. For now, it is best to consider 2019's relatively low rankings in some of the KPIs of BIMSTEC countries as indicators of areas for improvement rather than benchmarks of tourism competitiveness. As economies start recovering from the pandemic, tourism competitiveness will have to be reevaluated.

This report wherever possible uses illustrative examples of noteworthy activities, practices, and measures in each BIMSTEC country's tourism sector, especially those that could contribute to the establishment of thematic circuits. Highlighting one country's best practices is intended as an illustration of possible actions and should not be interpreted as suggesting that other BIMSTEC members are not implementing similar activities. The Secretariat has an important role to play in fostering an exchange of these activities among member countries, so they can learn from each other to increase regional cooperation that benefits all members.

---

[7]   World Economic Forum. Travel and Tourism Competitiveness Index. 2019. Geneva. https://www.weforum.org/reports/the-travel-tourism-competitiveness-report-2019/
[8]   In May 2022, WEF replaced the Index with the The Travel & Tourism Development Index, which expanded the KPIs with more emphasis on sustainability.

Situation Analysis of Global and Asia and the Pacific Tourism

Pre-COVID-19 Trends and Demand Drivers

Travel and tourism is one of the world's largest, fastest growing, and most resilient industries having contributed to global GDP a total of $9.2 trillion in direct, indirect, and induced contributions in 2019. Figure 1 shows the direct economic impact of the industry, including but not limited to accommodation, transportation, entertainment, and attractions, was $2.8 trillion in 2019.[9] Much of this was generated by the 1.5 billion international arrivals.[10] International arrivals grew 135% from 1996 to 2019 and were forecasted by the World Travel & Tourism Council (WTTC) to continue growing to 2.2 billion arrivals by 2029—that was until COVID-19 hit, reducing international travel practically to zero by April 2020.[11] As of August 2021, the recovery in international travel was still limited, but some countries reported growth in domestic and local travel, especially during the northern hemisphere summer months of July and August in both 2020 and 2021. Assuming a return to past growth, this could bode well for BIMSTEC in the longer-term and reaffirm the resilience of travel and tourism. But this will undoubtedly be a gradual process conditional on each country's COVID-19 situation, the COVID-19 situation in tourism generating markets, and the progress of vaccination.

Table 2 shows that pre-COVID-19 Asia and the Pacific was one of the world's fastest-growing regions for international tourist arrivals and one of the most resilient to shocks such as disasters triggered by natural hazards. International arrivals rose from 208 million in 2010 to 360.4 million in 2019, with a growth of 73%. Much of this growth was led by Northeast Asia (111.5 million arrivals in 2010; 170.3 million in 2019). Southeast Asia also, grew substantially, from 70.5 million to 138.6 million, as did some other subregions, but with fewer arrivals—Oceania from 11.5 million to 17.5 million and South Asia from 14.7 million to 34.0 million. International receipts in Asia and the Pacific rose from $370.8 billion in 2016 to $441.3 billion in 2019. Receipts in this period grew in all subregions, except Northeast Asia, down by 3.6%. The growth in the other subregions, however, reaffirms the industry's overall resilience. Despite political crises, health emergencies, and disasters triggered by natural hazards, the industry has shown continual resilience and always returned to growth.[12]

The substantial growth in tourism across Asia and the Pacific in 2019 generated 185 million jobs (10% of total jobs); over 79 million (4.5%) were directly generated by the industry.[13]

---

9   WTTC. Online Data Tool; WTTC Economic Impact Reports (accessed 20 August 2021).
10 UNWTO. World Tourism Barometer. July 2021.
11 WTTC. 2019. *Economic Impact Report*. London.
12 Data in this paragraph comes from the UNWTO World Tourism Barometer, July 2021 and December 2020.
13 WTTC. 2021. *Asia-Pacific 2021 Annual Research: Key Highlights*. London.

Figure 1: Global Tourism Gross Domestic Product Growth
($ billion)

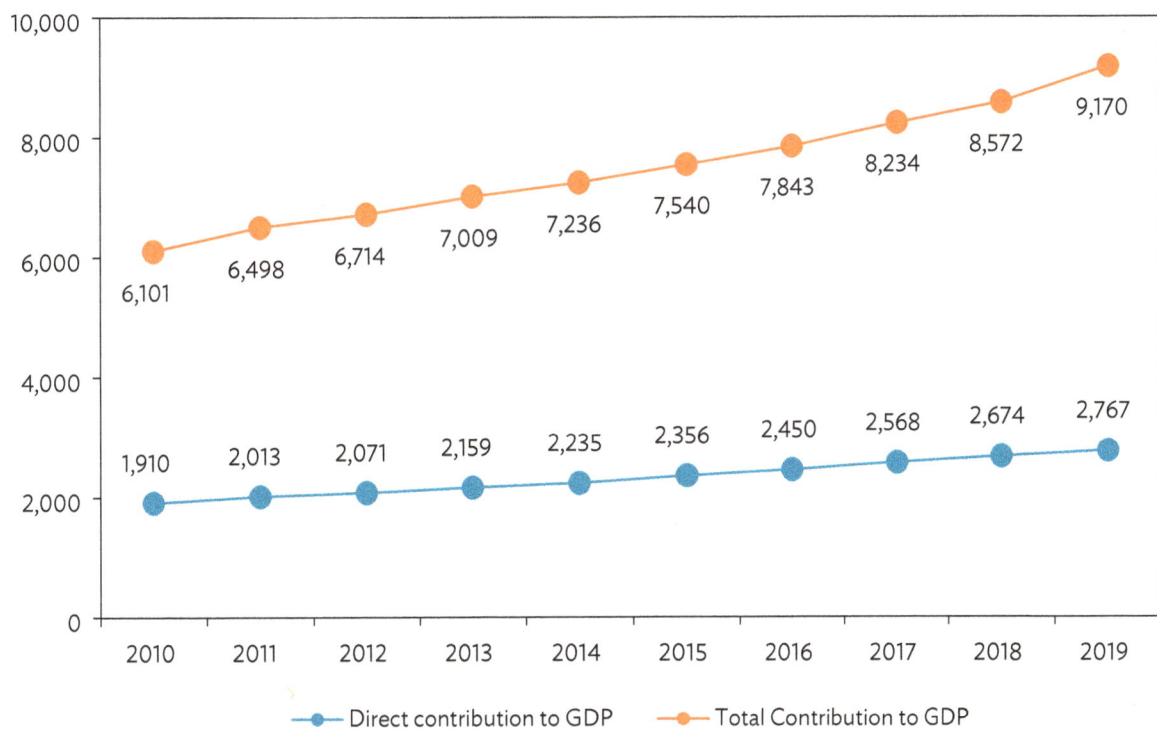

GDP = gross domestic product.

Source: World Travel & Tourism Council. Online Data Tool. Http:// tool.wttc.org (accessed 2 December 2020); WTTC Economic Impact Reports (accessed 20 August 2021).

Table 2: International Tourist Arrivals in Asia and the Pacific, 2010–2019

Region	2010	2015	2016	2017	2018	2019	Global Market Share (%)	Change 2019 vs 2018 (%)
APAC	208.0	284.0	305.0	323.3	346.5	360.4	24.7	1.0
NE Asia	111.5	142.0	154.0	159.5	169.2	170.3	11.7	(3.6)
SE Asia	70.5	104.0	110.8	120.6	128.6	138.6	9.5	3.8
Oceania	11.5	14.3	15.6	16.6	17.0	17.5	1.2	5.6
South Asia	14.7	23.5	25.0	26.7	31.7	34.0	2.3	6.2

( ) = negative, APAC = Asia and the Pacific, NE = Northeast, SE = Southeast.

Source: UN World Tourism Organization. World Tourism Barometer. July 2021 and December 2020.

As Figures 2 and 3 show, some of the fastest growing and most resilient countries in terms of international arrivals were, as of 2019, in South Asia and Southeast Asia. Myanmar topped the list with arrivals growth of almost 23% over 2018, followed by Bangladesh (21%), the Lao People's Democratic Republic (Lao PDR), and Viet Nam (16% each). International arrivals to Sri Lanka declined by 18% in 2019 because of the 2019 Easter bombings. Tables 3 and Table 4 show the number of international arrivals to BIMSTEC countries (minus Thailand) and Southeast Asia countries (with Thailand).

Figure 2: Percentage Growth in South Asia International Tourist Arrivals, 2018 and 2019

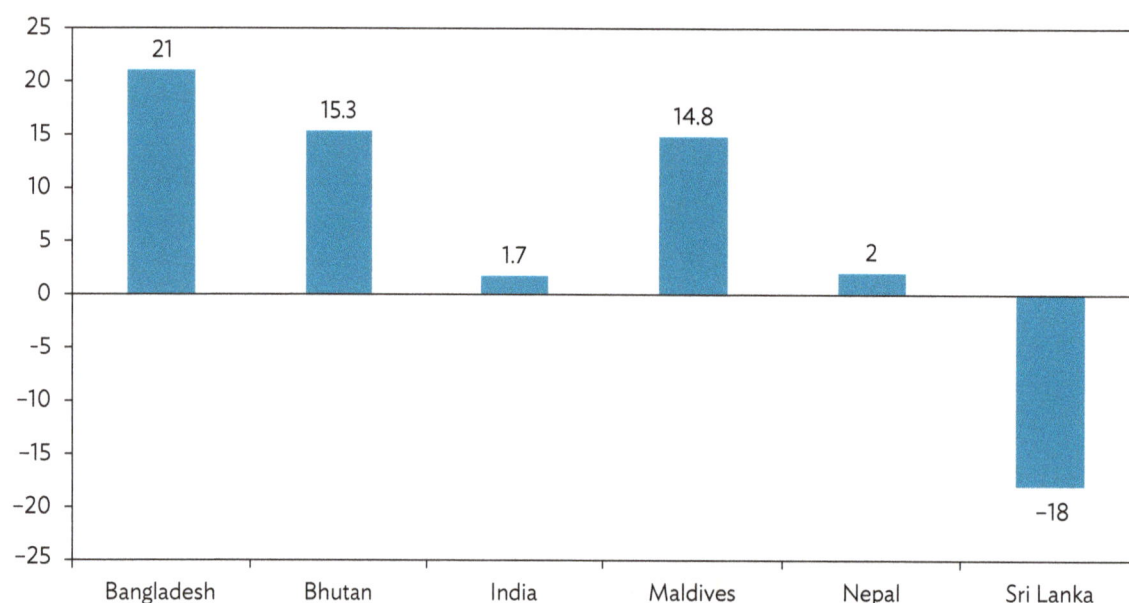

Source: UN World Tourism Organization. World Tourism Barometer. July 2021.

Table 3: International Tourist Arrivals in BIMSTEC Member Countries, 2018 and 2019
('000)

Country	2018	2019
Bangladesh	267	323
Bhutan	274	316
India	17,427	17,910
Myanmar	3,551	4,364
Nepal	1,173	1,197
Sri Lanka	2,334	1,914

Note: Data for Bangladesh were unavailable through UN World Tourism Organization. The source for 2018 and 2019 data was the Special Branch of the Bangladesh Police.

Source: UN World Tourism Organization. World Tourism Barometers. May 2021 and July 2021.

Figure 3: Growth in Southeast Asia International Tourist Arrivals, 2018 and 2019 (%)

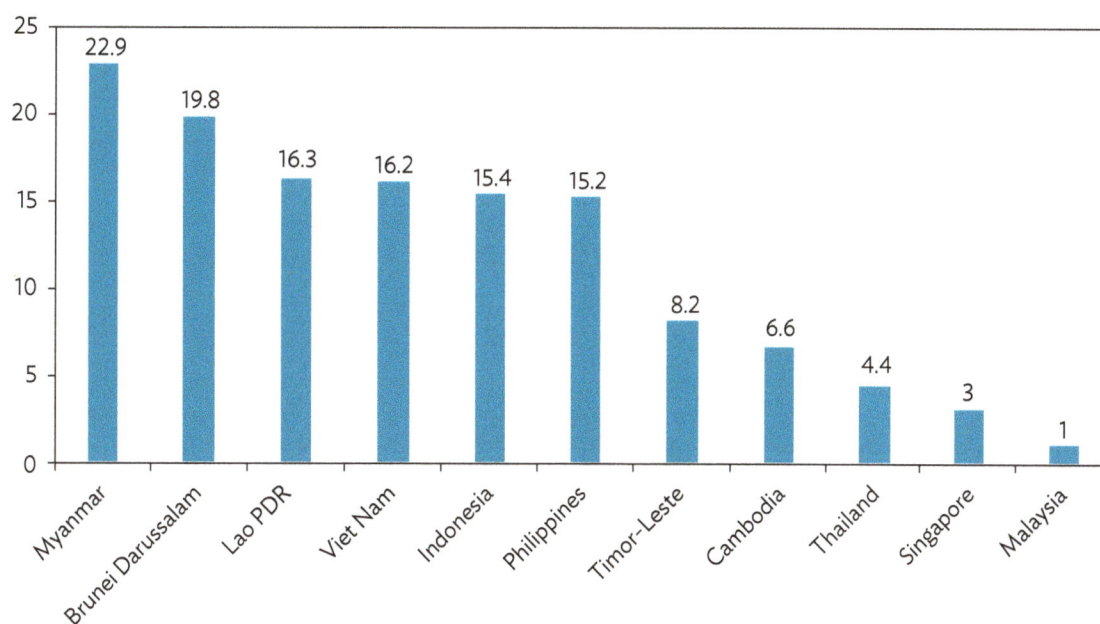

Lao PDR = Lao People's Democratic Republic.

Source: UN World Tourism Organization. World Tourism Barometer. July 2021.

Table 4: International Tourist Arrivals in Southeast Asia, 2018 and 2019 ('000)

Country	2018	2019
Brunei Darussalam	278	333
Cambodia	6,201	6611
Indonesia	13,395	15,455
Lao PDR	3,770	4384
Malaysia	25,832	26,101
Myanmar	3,551	4,364
Philippines	7,168	8,261
Singapore	14,673	15,119
Thailand	38,277	39,874
Timor-Leste	75	81
Viet Nam	15,498	18,009

Lao PDR = Lao People's Democratic Republic.

Source: UN World Tourism Organization. World Tourism Barometer. July 2021.

Main Drivers of Global and Asia and the Pacific Tourism

Before COVID-19 hit, the main demand drivers both globally and among Asia and the Pacific economies showed the resilience of tourism for BIMSTEC countries. The segments shown in Figure 4 are either already strong or offer strong potential for BIMSTEC countries, again keeping in mind that these are pre-COVID-19 drivers.

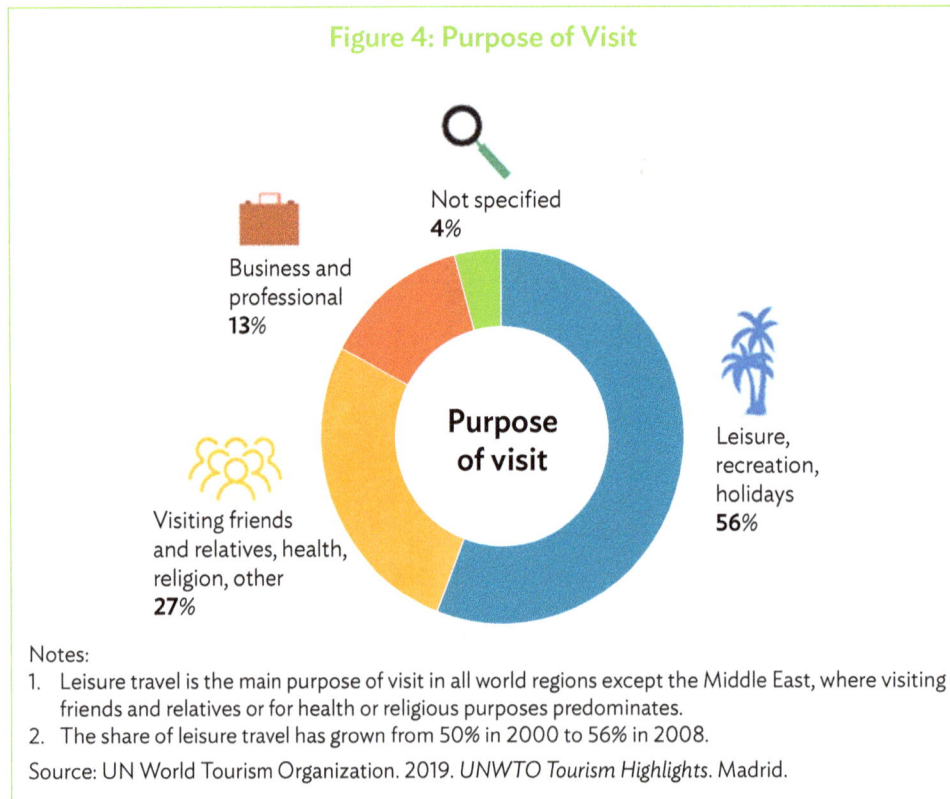

Figure 4: Purpose of Visit

Not specified
4%

Business and professional
13%

Purpose of visit

Leisure, recreation, holidays
56%

Visiting friends and relatives, health, religion, other
27%

Notes:
1. Leisure travel is the main purpose of visit in all world regions except the Middle East, where visiting friends and relatives or for health or religious purposes predominates.
2. The share of leisure travel has grown from 50% in 2000 to 56% in 2008.

Source: UN World Tourism Organization. 2019. *UNWTO Tourism Highlights*. Madrid.

In 2018, over half of the world's travel and tourism was driven by people seeking leisure, recreation, and holidays (56%), according to UNWTO's Tourism Highlights.[14] These also show that 27% were traveling for visits with friends and relatives, religion, and health—the percentage for this was higher in the Middle East. Some 13% traveled for business and the rest for other purposes (Figure 4). Before COVID-19, international tourist arrivals had been growing 4%–6% annually for decades, underpinned by rising disposable incomes, demand, and better connectivity—all demonstrating the industry's long-standing resilience.

These global drivers are also applicable to Asia and the Pacific. As the UNWTO explained in the *UNWTO/GTERC Asia Tourism Trends 2020 Edition*, 57% of international tourists in Asia and the Pacific traveled for leisure, recreation, and holidays, with other segments having similar percentages to global demand: visiting friends and relatives; trips for health, religion, and other reasons (25%); and business and professional travel (10%). The percentage of leisure and recreation travel in Asia and the Pacific grew from 48% in 2010 to 57% in 2019.[15]

---

[14]   UNWTO. 2019. *UNWTO Tourism Highlights*. 2019. Madrid.
[15]   UNWTO. 2020. *UNWTO/GTERC Asia Tourism Trends 2020 Edition*. Madrid.

Among the 56% traveling globally for leisure and recreation, beach vacations—whether to oceans, rivers, or lakes—were the chief attractions for approximately half of this group. An increasing number of international tourists to Asia and the Pacific sought more "authentic" destination experiences, such as nature-based and cultural attractions and activities, and entertainment. Increased interest in authentic experiences bodes well for the future of tourism in BIMSTEC countries, which have many of these experiences to offer via thematic circuits.

A 2019 Euromonitor International report lists 20 of the most influential megatrends set to influence the world.[16] Among these trends, the following influence the travel and tourism industry:

- healthy living, which includes an increased focus on wellness and the sharing economy (e.g., Airbnb and Uber);
- ethical living, including striving for and the preservation of authenticity;
- connected consumers, which is both the result and cause of increased multiculturalism and personalization; and
- Shopping reinvented, manifested by the fast global growth in online shopping.

These megatrends also reflect increased interest in and demand for sustainable products and services that respect and strengthen environmental and sociocultural protection.

For travel and tourism, these megatrends are reflected in the following drivers, which are already offered in all BIMSTEC countries before the COVID-19 pandemic and could be the basis for a circuit:

- **Adventure tourism.** The Adventure Travel Trade Association (ATTA) defines adventure tourism as a trip that includes at least two of the following three elements: physical activity, natural environment, and cultural immersion. Adventure activities are categorized as either "hard" or "soft," with the former tending to be more physically active. Hard activities include climbing and trekking, which is relevant for the Himalaya circuit between Nepal and Bhutan. Soft adventure activities include: backpacking, birdwatching, camping, canoeing, cycling, ecotourism, fishing, hiking, kayaking, rafting, and surfing. ATTA values adventure experiences at $683 billion, with the sector growing at a 21% compound annual growth rate since 2012.[17]
- **Cultural tourism.** This is one of the largest and fastest-evolving markets with an estimated four out of 10 tourists choosing their destination based on cultural offerings, according to the UNWTO.[18] It includes a wide range of activities, including pilgrimages, museum and cultural monument visits, and arts activities. It overlaps with culinary travel.
- **Gastronomy or culinary travel.** This is a cultural tourism segment and considered a soft adventure segment by ATTA. The World Food Travel Association forecasts substantial growth in food and wine festivals, and active combination experiences, such as culinary bike tours. Its research shows that most travelers spend at least 25% of their travel budgets on food and beverage.[19]

16   Euromonitor International. 2019. *Understanding the Socioeconomic Drivers of Megatrends.* p. 2. London.
17   Adventure Travel Trade Association. 2018. *20 Adventure Travel Trends to Watch in 2018.* Monroe, Washington. p. 5.
18   UNWTO. 2018. *Tourism and Culture Synergies.* Madrid, p. 74.
19   E. Wolf et al. 2020. *2020 State of the Food Travel Industry Report.*

- **Rural and community-based tourism.** The UNWTO says rural tourism is a "broader term that covers all forms of tourism usually practiced in rural areas, of which community-based tourism is one." Community-based tourism (CBT) comprises "forms of tourism that involve high engagement with the communities encountered." CBT aims to benefit communities, as well as enrich and preserve local traditions and culture.[20] CBT can also involve adventure, cultural, and culinary tourism. BIMSTEC is rich in potential and actual CBT experiences. Circuits can be effective ways to connect and engage communities through tourism.
- **Wellness tourism.** Wellness is becoming a high priority, globally and particularly in Asia. The Global Wellness Institute defines wellness as the "active pursuit of activities, choices, and lifestyles that lead to a state of holistic health."[21] It defines wellness tourism as "travel associated with the goal of maintaining or enhancing one's personal well-being and includes the pursuit of physical, mental, spiritual or environmental 'wellness' while traveling for either leisure or business."[22] The wellness tourism market is an increasingly prominent driver, accounting for an estimated total (direct, indirect, and induced) spending of $683 billion in 2018, according to the Global Wellness Institute. Figure 5 shows that wellness is a diverse segment comprised of multiple activities and types of attractions. This segment is expected to do quite well after the COVID-19 pandemic passes, especially in Asia. This is because it is a region that prizes wellness and healthy living. Wellness is often associated with medical tourism, a significant segment for BIMSTEC. This segment specifically focuses on traveling for medical treatments and procedures, while wellness comprises a broader set of activities.

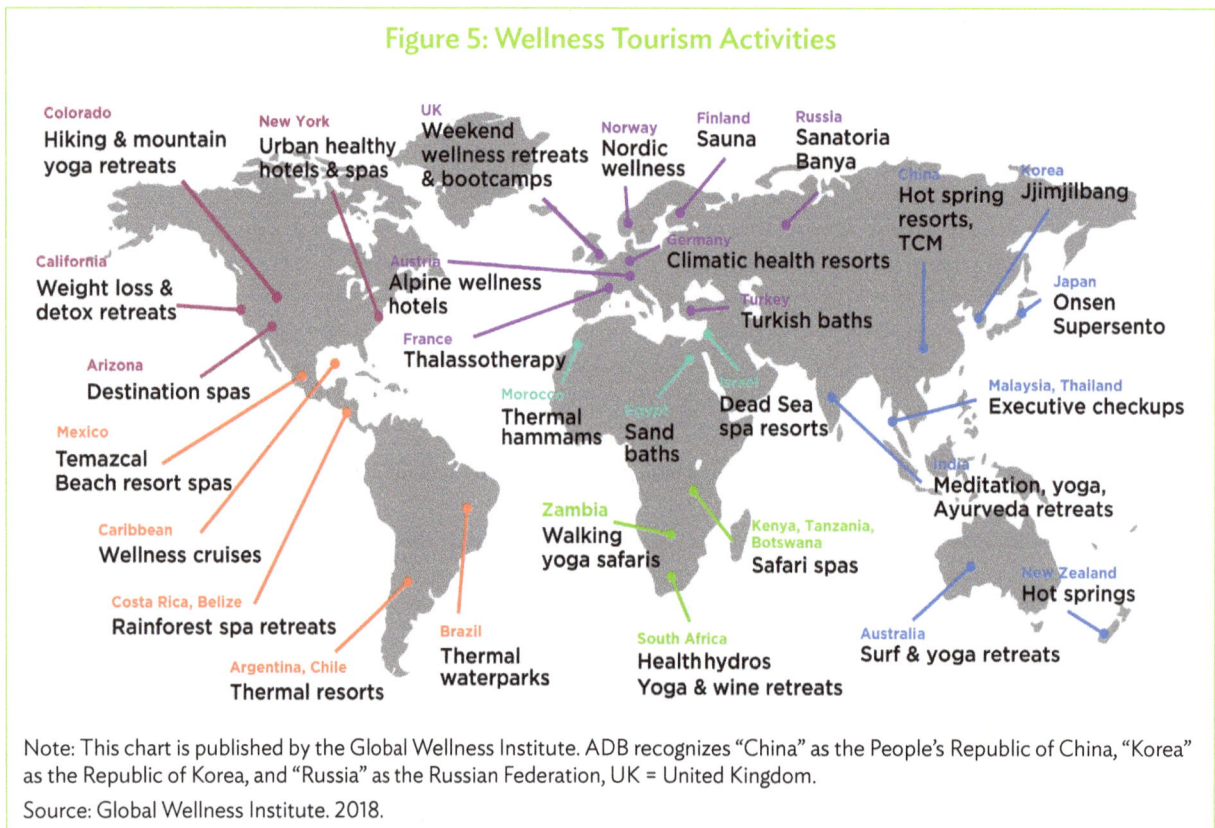

Figure 5: Wellness Tourism Activities

Note: This chart is published by the Global Wellness Institute. ADB recognizes "China" as the People's Republic of China, "Korea" as the Republic of Korea, and "Russia" as the Russian Federation, UK = United Kingdom.

Source: Global Wellness Institute. 2018.

20   UNWTO. 2020. *UNWTO/GTERC Asia Tourism Trends*. Madrid. p. 72.
21   Global Wellness Institute. 2018. Global Wellness Tourism Economy. Miami.
22   Global Wellness Institute. 2013. *Global Wellness Tourism Economy*.

Theme parks, water parks, and attractions. The world's top parks and attractions increased attendance by 4% in 2019 (521 million visitors). Twelve of the top 25 amusement parks are in Asia, which accounted for over 60% of global growth in this segment. That growth was expected to continue at 5.3% annually to 2022, but it was derailed by COVID-19. Operators based in the People's Republic of China (PRC), such as the Overseas Chinese Town Enterprises Co. (OCT Parks) and the Fantawild Group are two of the world's largest theme park operators, with the latter growing 9.4% in 2019, making it larger than the global Universal Parks & Resorts.[23]

India and Thailand are theme park destinations with at least 35 and 40 parks, respectively. India has announced plans to establish a Buddhist theme park at Futala Lake in Gujarat state at an estimated cost of $134 million. The park will be part of a Gujarat tourist circuit in the Saurashtra region that will include pilgrimage sites such as Somnath Jyotirlinga and Girnar.[24] A theme park is planned in Uttarakhand near the Shri Daksheshwar Mahadev Temple and 52 shrines and pilgrimage sites (Shakti Peethas) that are important to Shaktism, a Hindu sect.[25] Theme parks in Thailand include the Ramayana and Cartoon Network Amazone water parks.[26]

Theme parks and wellness tourism developments are often closely aligned and integrated with destination hotel developments. Hotel industry development trends are often interdependent with the development of attractions. The Buddhist theme park in India, for example, will attract hotel investors as it progresses, and the types of hotel investments that are attracted could determine tourist spending levels at the park. The magnitude of hotel investments can have substantial positive impacts on a destination. For context, globally, the hotel industry's market size was estimated at $1.21 trillion in 2019.[27] The market comprised 18 million rooms; 54% of them affiliated with a global or regional hotel chains up from 50% in 2012, and 46% unaffiliated. The top five hotel groups in terms of market size—IHG, Marriott, Hilton, Wyndham, and Accor—accounted for 25% of market share, up from 19% in 2012. These groups account for 58% of the global development pipeline of hotels (planned or under construction).[28]

Wellness is becoming an important consideration for the lodging industry. Hyatt, for example, has incorporated elements of wellness into its entire business and had named a senior vice-president for global wellbeing to lead this effort, Mia Kyricos. Kyricos told travel industry news site Skift that, "We view well-being as more than spas, fitness centers, and healthy food options ... [W]e care for people so they can be their best [with] ... wellness as the road and well-being as the destination."[29]

Given the size of these sector estimates, it is important to remember that they overlap, especially cultural tourism and adventure travel, both of which focus on providing authentic and memorable experiences with the latter emphasizing nature-based activities. Wellness tourism, in turn, depends on a healthy environment and many activities in this segment are related to natural products and nature-based activities. Wellness tourism is also often an integral part of resort and leisure-based offers.

Among the megatrends and drivers for travel and tourism, technology both enables and drives the innovations that are shaping these markets. Online content and booking aggregators, such as Expedia, Orbitz, and Priceline, have been in business for over 2 decades and helped make more destinations and experiences available to travelers

23 Themed Entertainment Association and Architecture, Engineering, Construction, Operations, and Management (TEA/AECOM). 2019. *Theme Index and Museum Index: The Global Attractions Attendance Report.* Los Angeles. p. 6.
24 Interpark. 2020. India investing heavily in theme parks and attraction sites. 10 August.
25 B. Mitchell. 2020. Religious theme park planned for Haridwar, India. *Blooloop.* 28 July 2020.
26 InterPark news articles on attractions in Thailand, 2016–2021.
27 Statista. 2021. Hotel Industry Worldwide. p. 2. Hamburg (accessed 28 November 2021).
28 Statista. 2021. Tourism Worldwide. pp. 12–16. Hamburg (accessed 28 November 2021).
29 Skift. Megatrends Defining Travel in 2019. New York.

worldwide. Hundreds of apps have been created to help with accommodation, transport, local tours, and even how to pack. Technology is significantly influencing accommodation and transportation providers to optimize sales, routings (i.e., low-cost carriers), service offers, and sustainability efforts.

Airbnb is a fast-growing technology-related player in lodging and is competing with mainstream lodging companies. Airbnb had over 7 million rental listings available in 220 countries and more than 100,000 cities in 2020.[30] The company is also offering more than 40,000 travel "experiences" worldwide. The top three cities in terms of the number of Airbnb bookings were Tokyo, Paris, and New York. Airbnb generated $4.81 billion in revenue in 2019, up 32% from 2018.[31] In 2016, Airbnb had 3 million bookings in total and 1 million listings comparable to hotel rooms. For comparison, Marriott had 1.1 million hotel room listings.[32] Hotels, however, have higher overall occupancy rates, average daily rates, and revenue per available room than Airbnb. Some cities are beginning to regulate Airbnb because the rapid increase in listings is causing market distortions, such as sharply rising housing rents.

International air transport spending grew from $173 billion in 2010 to $250 billion in 2019, comprising about 15% of all tourism exports, according to the UNWTO.[33] In 2019, according to the ForwardKeys research consultancy, air passenger journeys rose by 4.5%. Departures from Asia and the Pacific increased an estimated 7.7% in 2019, while intraregional air travel grew by 8.7% (footnote 33).

Demand for cruise travel has also been brisk, rising 20.5% over the last 5 years. Cruise Lines International Association reported that in 2018, 28.5 million passengers cruised globally, up by 6.7% from 2017. In 2016, the association estimated the industry had $126 billion in total economic impact and generated 1 million jobs. Although this is one of the fastest-growing segments of the leisure travel market, it still represents only 2% of the overall global travel and tourism industry.[34] Cruise passengers grew by 7% in 2018, which was even faster than the growth of international tourist arrivals at 6%, according to the UNWTO.[35]

All the demand drivers just discussed are global, but some of the fastest growth has been in Asia and the Pacific, which is not surprising since about a quarter of the world's tourism is concentrated in this region. Other drivers also contributed to this growth, including: [36]

- The Asia and the Pacific economy grew rapidly from 2000 to 2018, averaging 5% annual growth, accounting for $30 trillion or 35% of the world economy.
- The PRC was the world's largest outbound travel market, rising from 4.5 million travelers in 2000 to more than 150 million in 2019.
- Many economies in Asia and the Pacific received substantial numbers of visitors from the PRC before the pandemic.
- Per capita incomes in Asia and the Pacific economies increased on average 160% from $2,700 in 2000 to $7,000 in 2018, which, in turn, generated increased disposable income and interest in outbound travel.
- Shopping, sightseeing, and wellness are priority activities.

30   Airbnb News. 2021. Fast Facts.
31   Statista. 2021. Airbnb Dossier. p. 3. (Accessed November 28, 2021)
32   Marriott Our Story
33   UNWTO. 2020. World Tourism Barometer. January 2020.
34   Cruise Lines International Association. 2020. *2020 State of the Cruise Industry Outlook Report.*
35   UNWTO. World Tourism Barometer. January.
36   UNWTO. 2020. *UNWTO/GTERC Asia Tourism Trends.* Madrid. p. 72.

- International travel within Asia has been boosted by improved air connectivity, better infrastructure, and travel facilitation and promotion (noting over half the world's population lives in Asia and the Pacific).
- Intraregional tourism predominated in Asia and the Pacific: 79% of the region's international tourist arrivals originated in Asia in 2018.

Impact of COVID-19 on Global and Regional Travel

COVID-19 hit not long after the UNWTO released its January 2020 World Tourism Barometer, which highlighted the travel and tourism industry's positive performance over the past decade. By April, the UNWTO reported that 100% of all countries worldwide had totally or partially closed their borders. Figure 6 shows international arrivals sank to zero worldwide.

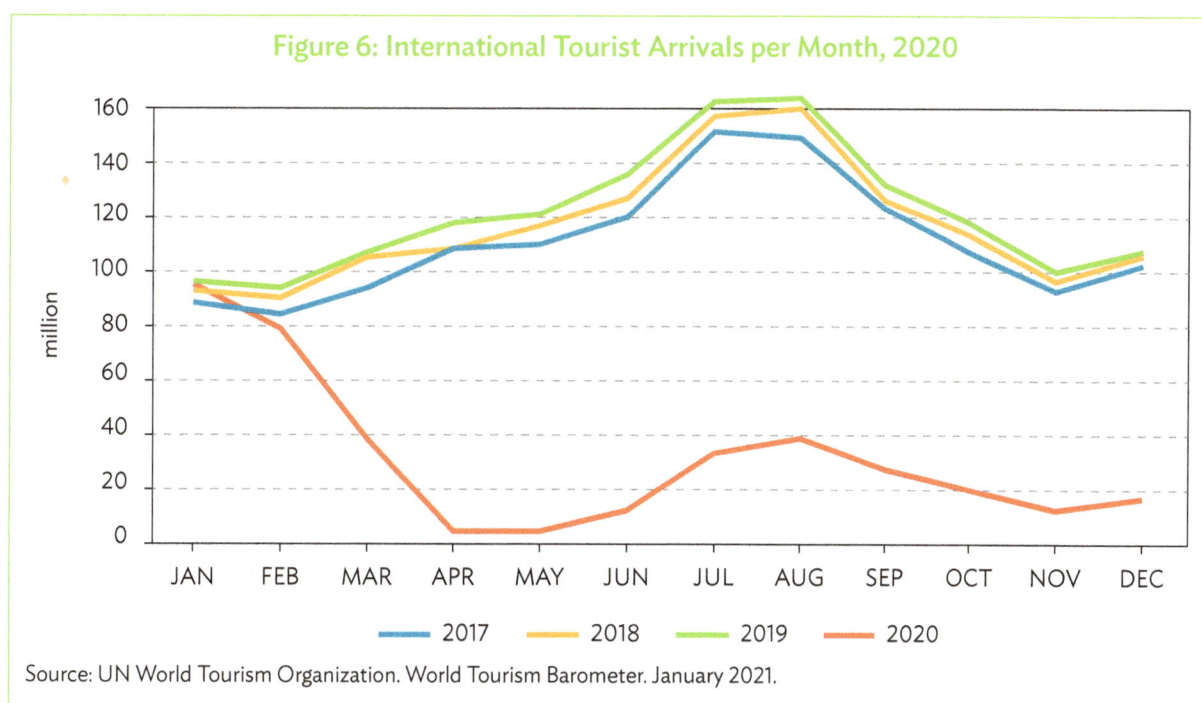

Figure 6: International Tourist Arrivals per Month, 2020

Source: UN World Tourism Organization. World Tourism Barometer. January 2021.

On every continent, travel and tourism stopped amid massive cancellations and the closure of tourism facilities and borders, causing millions of tourism- related job losses. While travel and tourism has consistently proven to be a resilient industry, the impact of COVID-19 has been severe. Many tourism-related businesses folded in 2020, and up to 62 million jobs in the tourism were reportedly lost globally.[37] Industry experts are optimistic that, with the global distribution of COVID-19 vaccines, travel and tourism will eventually return to 2019 levels. There will be pent-up demand, but the shape of what experts believe will be an altered tourism market is constantly changing. Because of COVID-19, travelers and the industry have become more conscious about health, hygiene, and safety. A debate is emerging on whether post-COVID-19 travel will be a return to past demand trends or whether new demands, particularly on health and hygiene, will shape the industry. A likely scenario is a hybrid of both whereby

---

[37] WTTC. 2021. *Travel & Tourism Economic Impact 2021*. London. p. 3.

destinations will need to demonstrate higher levels of sanitation, cleanliness, and health standards. More emphasis will need to be put on this in tourism strategies and plans for every destination.

The CDC tracks COVID-19 in all economies and rates them on risk levels from 1 (low) to 4 (very high). The CDC advises travelers to avoid all Level 4 destinations. As mentioned previously, the CDC in 2021 recommended against all travel to 74 economies, including one BIMSTEC country: Myanmar (Table 5). Sri Lanka was at a Level 3 high risk level (recommended travel only if vaccinated), India and Bhutan were at a Level 1 (low risk).[38] However, it is important to note these warning levels were changing every week in 2020–2021, depending on the level of COVID-19 spread and risk.

Table 5: COVID-19 Travel Recommendations by Destination

Level 4: COVID-19 Very High	Level 3: COVID-19 High	Level 2: COVID-19 Moderate	Level 1: COVID-19 Low	Level Unknown: COVID-19 Unknown
"Avoid travel to these destinations. If you must travel to these destinations, make sure you are fully vaccinated before travel."	"Make sure you are fully vaccinated before traveling to these destinations. Unvaccinated travelers should avoid nonessential travel to these destinations."	"Make sure you are fully vaccinated before traveling to these destinations. Unvaccinated travelers who are at increased risk for severe illness from COVID-19 should avoid nonessential travel to these destinations."	"Make sure you are fully vaccinated before travel to these destinations."	"Avoid travel to these destinations. If you must travel to these destinations, make sure you are fully vaccinated before travel."
Armenia Azerbaijan Brunei Darussalam Georgia Guam Malaysia Maldives Mongolia Myanmar New Caledonia Papua New Guinea Singapore	Australia Fiji French Polynesia Lao PDR Philippines Republic of Korea Sri Lanka Thailand Viet Nam	Nepal New Zealand	American Samoa Bangladesh Bhutan Hong Kong, China India Indonesia Japan Kyrgyz Republic Marshall Islands Micronesia, Federated States of Northern Mariana Islands Palau Pakistan People's Republic of China Saba Taipei,China Timor-Leste	Cambodia Cook Islands Democratic People's Republic of Korea Kazakhstan Kiribati Macau, China Nauru Niue Samoa Solomon Islands Tajikistan Tokelau Tonga Turkmenistan Tuvalu Uzbekistan Vanuatu Wake Island

Lao PDR = Lao People's Democratic Republic.

Source: Centers for Disease Control and Prevention. COVID-19 Travel Recommendations by Destination (accessed 6 December 2021).

---

[38]   Centers for Disease Control and Prevention. COVID-19 Travel Recommendations by Destination (accessed 20 August 2021).

The COVID-19 travel regulations of the International Air Transport Association (IATA) indicated that as of December 2021 every country in the Asia-Pacific region was "partially restrictive"," except Lao PDR and Myanmar, which were indicated as "totally restrictive." IATA does not define the difference between the partially and totally restrictive designations.[39]

The airline industry was especially hit hard by COVID-19 with traffic having dropped precipitously as the overall tourism industry stopped. But some tourism markets began to reopen in December 2020 in destinations where COVID-19 was a reduced threat. In the EU, travel reopened between some countries in the bloc. But this situation remains highly fluid and is determined by COVID-19 threat levels. Figure 7 shows global air capacity as of 20 August 2021. Some markets have grown since January 2020, including the PRC, the Russian Federation, Spain, Türkiye, and Greece, but the other 15 markets in the figure are still at lower capacity.

Figure 7: Reductions in Scheduled Air Capacity in Main Tourism Generating Markets

Country	15-Jul-19	20-Jan-20	05-Jul-21	12-Jul-21	% Change Week on Week	% Change V's 20th Jan 2020	% Change V's W/C 15th Jul' 2019
United States	23,570,596	20,749,829	19,389,546	19,556,224	0.9	−5.8	−17.0
PRC	16,959,400	16,882,726	17,324,943	18,140,436	4.7	7.4	7.0
Russian Federation	2,767,383	2,158,058	2,781,979	2,759,240	−0.8	27.9	−0.3
Spain	3,682,634	2,226,308	2,336,274	2,452,005	5.0	10.1	−33.4
India	4,002,429	4,255,510	2,255,799	2,397,079	6.3	−43.7	−40.1
Türkiye	2,630,682	1,924,284	2,078,706	2,180,288	4.9	13.3	−17.1
Japan	4,234,858	4,121,355	1,789,768	1,881,366	5.1	−54.4	−55.6
Italy	2,731,605	1,775,401	1,697,879	1,768,701	4.2	−0.4	−35.3
Mexico	1,946,123	1,920,941	1,682,178	1,727,423	2.7	−10.1	−11.2
Brazil	2,649,847	2,842,645	1,679,653	1,724,983	2.7	−39.3	−34.9
France	2,596,224	1,842,023	1,593,191	1,602,698	0.6	−13.0	−38.3
Germany	3,363,020	2,519,489	1,519,500	1,558,081	2.5	−38.2	−53.7
United Kingdom	3,892,659	2,712,915	1,048,700	1,172,977	11.9	−56.8	−69.9
Korea, Republic of	1,814,522	1,823,750	972,418	973,353	0.1	−46.6	−46.4
Greece	1,159,765	319,763	929,938	968,378	4.1	202.8	−16.5
Australia	2,146,057	2,058,708	1,095,635	928,973	−15.2	−54.9	−56.7
Indonesia	2,963,074	3,144,407	1,740,561	895,481	−48.6	−71.5	−69.8
Saudi Arabia	1,296,017	1,344,926	804,484	831,893	3.4	−38.1	−35.8
Canada	2,275,514	1,884,093	749,251	788,227	5.2	−58.2	−65.4
Colombia	896,088	895,513	705,240	701,903	−0.5	−21.6	−21.7

PRC = People's Republic of China.

Source: Overseas Airline Guide Blog (accessed 20 August 2021). Table is reproduced directly from OAG.

Scheduled flights from and within the PRC were up 7.4% as of 12 July 2021 from January 2020—a potentially positive sign for BIMSTEC since the PRC is one of the largest generating, spending, and receiving markets for member countries.

Future of Global Travel and Tourism

Until COVID-19 vaccines are widely distributed globally, international travel will remain restricted as countries continue to implement maximum health, hygiene, and safety protocols. Various guidelines, protocols, and approaches were tried or tested for travel and tourism in 2020 and 2021, including:

---

[39]   International Air Transport Association. (accessed 6 December 2021). The levels of restrictions change on a daily basis.

- Many countries focusing on domestic, local, and day-visit markets until international markets reopen.
- Travel being allowed between destinations determined to be safe. An example of publicizing these destinations is the European Union's Re-open EU website and app.[40] In November 2020, the ASEAN Declaration on an ASEAN Travel Corridor Arrangement Framework was adopted at an online summit. The framework offers a possible model for BIMSTEC and is described in Section 4.
- The UNWTO and WTTC introduced the Safe Travels global protocols. The WTTC has specified protocols for nine industry segments.[41]
- Many countries increased their public awareness campaigns to maintain health and safety measures at destinations and communicated measures to visitors.

The sudden decline of the industry resulted in many travel and tourism-related businesses closing or going bankrupt, the loss of tens of millions of jobs, and billions of dollars in lost income and revenue. In its 2021 Economic Impact Report, WTTC estimated that as many as 62 million travel and tourism jobs were lost due to the pandemic, many of which were in SMEs with limited resources.[42] Travel and tourism will restart first in those destinations that are best prepared to control COVID-19, vaccinate their populations, and implement the measures necessary to ensure public health and safety. This will influence where people travel and what they do at their destinations. Both travelers and service providers—tour operators, airlines, cruise lines, hotels, and restaurants, among others—will need to be reassured of this.

International cross-border travel, such as for thematic circuits, will be limited, if not impossible, until destinations can reduce the spread of COVID-19 and vaccinate large numbers of people living in destination areas. If this is achieved, international travel will resume—and here thematic circuits might offer a faster track path for the industry's recovery. For those BIMSTEC countries with improved health and safety conditions and control over COVID-19, a broad range of cross-border thematic circuits could be on offer, as Section 2 explains.

Why Thematic Circuits?

The UNWTO defines a tourist circuit as a "subsystem of tourism that focuses on the needs of people while geographically mobile, including transport facilities, availability of information, and the proximity of stopover attractions. From the standpoint of the tourist, the act or experience of briefly visiting a number of areas as part of a single round trip."[43]

In short, a circuit is an organized tourist itinerary that facilitates visits to a series of destinations, which are often connected by one or more cultural, historical, or religious themes. A circuit enables several destinations in an area to collaborate on development and marketing, thus increasing their attractiveness to visitors and tour operators, as well as investors. Circuits also increase opportunities for community engagement at destinations offering authentic products, visitor experiences, and services to visitors, possibly all year. Greater community engagement in urban and rural areas could and increase the participation of microenterprises, thereby generating employment opportunities, especially for youth and women. The UNWTO cautions, however, that this needs to be handled carefully since tourism can bring social changes to communities and possibly have impacts on women.[44] But with the careful consideration of the potential negative impacts, cross-border circuits can amplify the positive effects of

---

40   https://reopen.europa.eu/en.
41   WTTC. WTTC Safe Travels Protocols.
42   WTTC. 2021. *Travel & Tourism Economic Impact 2021*. London. p. 3.
43   UNWTO and Secretariat of State for Tourism of France. 2001. *Thesaurus on Tourism and Leisure Activities*. p. 373.
44   UNWTO. 2020. *UNWTO/GTERC Asia Tourism Trends*. p. 91. More research is recommended on the implications and opportunities for women and circuit development.

community engagement. Indeed, doing this for itineraries that span two or more countries could increase cultural understanding and help promote regional integration.

The plethora of online information and booking channels accessible to travelers and tour operators means that itineraries—and thus circuits—can easily be created by either individuals or tour operators. For Spain's Camino de Santiago, for example, the local government provides an informative interactive website that aggregates and presents a broad range of visitor information for trip planning. For the Buddhist circuit, a visitor wanting to travel in the footsteps of Buddha can follow the detailed information in *Lonely Planet* guidebooks for India and Nepal or book a tour with the many local or international operators that offer Buddhist circuit tours.

Governments have an important role to play in establishing thematic circuits. Governments individually and jointly through regional cooperation can enrich and facilitate the visitor experience by providing information and helping to provide services and investments that ordinarily could not be easily provided—and sometimes not at all—by the private sector. These include infrastructure needs such as roads and smoother border-crossing, destination-wide marketing and product development, visitor information, and sector-wide education and training, as well as policies such as relaxed visa requirements, investment incentives, and increased air access via Open Skies agreements. As destinations struggle to recover from COVID-19, it is more important than ever for governments to establish and enforce international best practices on health, hygiene, and safety protocols and standards.

The private sector depends on government to provide the infrastructure for circuits and marketing and promotion support for destinations. In parallel, the tourism value chain of services provided by the private sector is essential for putting together itineraries and operationalizing circuits. If destinations along an existing or prospective circuit offer interesting visitor experiences and attractions, then tour operators will move forward with tour offers and itineraries that they can market. Adventure travel is an exception because rough roads and a less developed destination are part of the experience.

A circuit can be considered successful and sustainable when it does the following:

- It contributes to preserving and protecting cultural and natural heritage attractions and surrounding areas. Preservation also reinforces the authenticity of heritage offers.
- Provides interesting attractions and experiences for visitors and tour operators, while extending lengths of stay and increasing spending.
- It provides potentially lucrative opportunities for investors and businesses, either directly or on a public–private partnership (PPP) basis.
- It benefits local-destination communities by facilitating the creation of microenterprises and increasing employment, especially for women and youth.

A prime example of a destination in demand along a circuit is the city of Bodh Gaya in the Indian state of Bihar, which is considered the birthplace of Buddhism and thus central to the Buddhist circuit. This circuit would benefit, however, from improvements in the city's infrastructure, especially a sewage system, adequate water supply, and international-level hotels. Despite this, in 2019 Bodh Gaya still attracted over 1.3 million domestic and nearly 254,000 international visitors, many of them joining one of several circuit itineraries offered by hundreds of tour operators.[45] The number of international visitors, however, is a tiny fraction of the potential international market of over 400 million Buddhists.

---

45 M. Bhonsale. 2019. *Religious Tourism as Soft Power: Strengthening India's Outreach to Southeast Asia. ORF Special Report No. 97*. Observer Research Foundation. September.

Some of the world's best-known circuits include:

- Camino de Santiago
  - This circuit is a large network of ancient pilgrimage routes across Europe converging in northwest Spain, one of which stretches 780 kilometers (km) dating to the 9th century (Figure 8). The circuit attracted more than 300,000 visitors in 2017, many of them walking long segments of the routes. Pilgrims can buy a special "passport," which is stamped at each stop along the circuit. United Nations Educational, Scientific and Cultural Organization made it a World Heritage Site in 1993.
  - The tourism office of the Galicia regional government offers seven different sets of visitor experiences, which include itineraries of 4–8 days for tours by bike, horseback, sailboat, and off-road vehicles. The government's website also includes interactive maps with information on accommodation, dining, things to see and do, and transport. A useful Build Your Trip app enables visitors to select trip components and build their itineraries.

Figure 8: Camino de Santiago Build Your Trip App

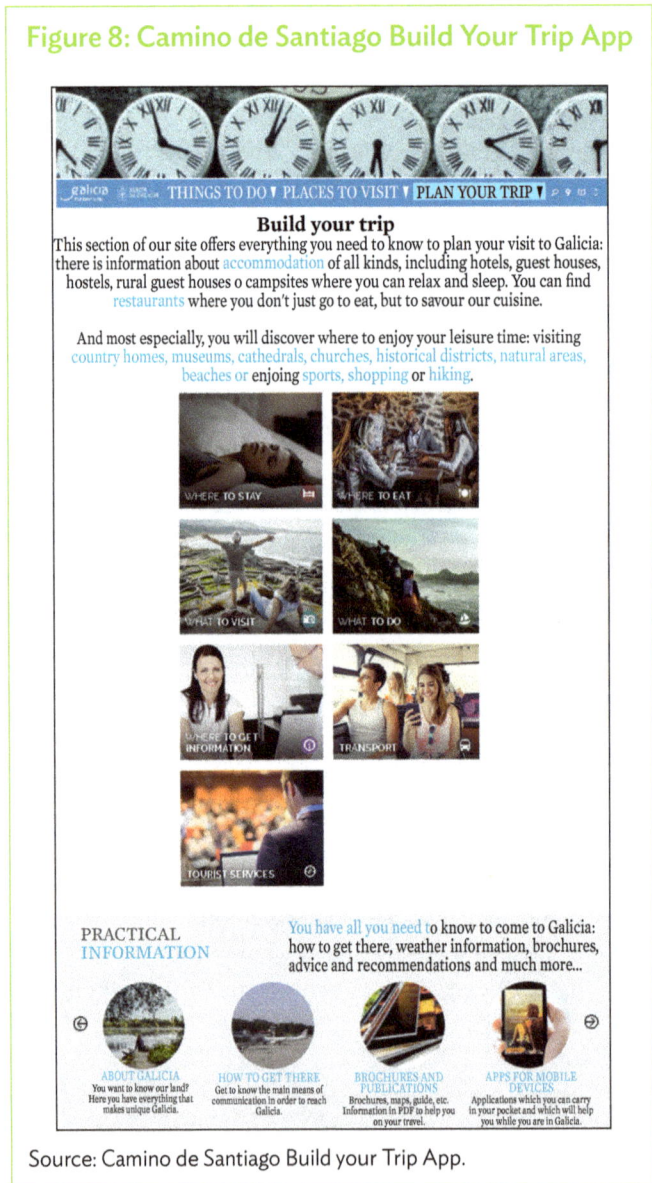

Source: Camino de Santiago Build your Trip App.

- The Buddhist circuit
  - This circuit is a focus for further development in Nepal and especially the Indian states of Uttar Pradesh and Bihar. The circuit follows in the footsteps of the Buddha across the Gangetic Plains from Lumbini in Nepal, where he was believed to have been born, through Bihar, where he believed to attained enlightenment, to Sarnath and Kushinagar in Uttar Pradesh, where he believed to gave his first teachings and died.
  - The state governments of Uttar Pradesh and Bihar have visitor information websites with descriptions of tourist sites for the Buddhist circuit. Bihar also has information on the Eco circuit, the Jain circuit, the Ramayana circuit, the Sikh circuit, and the Sufi circuit.
  - Many tour operators in BIMSTEC countries, as well as outside the region, offer Buddhist circuit tours. An online search on Viator.com, for example, shows 545 Buddhist circuit tours of varying durations.[46]

- The Silk Road
  - This circuit connects the 34 countries that consider themselves part of the ancient Silk Road stretching from Asia to the Middle East and into southern Europe. A lesson learned in promoting this circuit was the need for a coordinating entity to select activities and help implement them. The UNWTO set up the Silk Road Task Force to assist its members via workshops for marketing and promotion, capacity building and destination management, tourism route development, and research. The Chang'an–Tianshan Corridor, a 5,000-km section of the Silk Road, is a UNESCO World Heritage Site.

Additional thematic circuits for consideration include those highlighted in the Greater Mekong Subregion Tourism Strategy, 2016–2025 (GMS):

- **The Southern Coastal Corridor.** Cambodia, Myanmar, Thailand, and Viet Nam for beach and islands, leisure, seafood, history, culture, and community-based tourism.
- **The Mekong Tea Caravan Trail.** The PRC's Yunnan Province, the Lao PDR, Myanmar, and Thailand for ethnic group experiences, culture, and ecotourism.

The GMS Strategy also promotes multi-country sporting events—something BIMSTEC could consider. Events include the annual Mekong Triathlon, cross-border marathons and a possible Tour de Mekong cycling race.

The UNWTO at its 98th Executive Council session in June 2014 discussed the role of tourism routes for fostering regional development and integration. A summary report of this discussion drew attention to the benefits of themed routes and success factors and challenges for their development. This report could be helpful for creating BIMSTEC circuits and includes:[47]

- strategies for integrated product development, innovation, and new technologies;
- the creation of "route" brands; and
- governance models, including PPPs, public-to-public coordination, and the creation of networks.

These factors are considered in the following section that examines the issues facing the overall development of tourism, especially circuits, in BIMSTEC countries.

---

[46] Buddhist circuit tours listed on https://www.viator.com/searchResults/all?text=Buddhist+Circuit.
[47] UNWTO Executive Council, Session 98, Thematic discussion: The Role of Tourism Routes in Fostering Regional Development and Integration. 2 May 2014. Santiago de Compostela.

Overall Tourism Trends in the Bay of Bengal Initiative for Multi-Sectoral Technical and Economic Cooperation Region

Pre-COVID-19 Trends

International tourism trends were moving in a favorable direction for all BIMSTEC countries before COVID-19 hit. According to latest data from the UNWTO and the WTTC, the industry was having a meaningfully positive economic impact across the region. Table 6 shows growth in arrivals and receipts since 2000. And that favorable trend continued to until February 2020 in most of these countries. That said, Myanmar, Nepal, and Sri Lanka experienced a few declines during this period due to local factors.

Table 6: BIMSTEC International Tourist Arrivals and Receipts

BIMSTEC Member Countries	2000	2005	2010	2015	2016	2017	2018	2019
International Tourist Arrivals ('000)								
Bangladesh	199	208	303	643	830	237	267	323
Bhutan	8	14	41	155	210	255	274	316
India	2,649	3,919	5,776	13,284	14,570	15,543	17,427	17,910
Myanmar	208	660	792	4,681	2,907	3,443	3,551	4,364
Nepal	464	375	603	539	753	940	1,173	1,197
Sri Lanka	400	549	654	1,798	2,051	2,116	2,334	1,914
Thailand	9,579	11,567	15,936	29,923	32,529	35,591	38,178	39,916
Total	**13,507**	**17,292**	**24,105**	**51,023**	**53,851**	**58,806**	**63,303**	**65,256**
International Tourism Receipts ($ million)								
Bangladesh	50	75	81	150	214	337	353	388
Bhutan	10	19	40	71	73.7	80	85.4	89
India	3,460	7,493	14,490	21,013	22,427	27,365	28,568	29,962
Myanmar	162	67	72	2,120	2,197	1,969	1,652	2,483
Nepal	158	132	344	481	446	630	641	701
Sri Lanka	253	362	576	2,981	3,518	3,925	4,380	3,607
Thailand	7,483	9,576	20,104	44,922	46,274	53,951	58,066	61,571
Total	**13,576**	**19,729**	**37,717**	**73,753**	**77,166**	**90,274**	**93,224**	**98,801**

BIMSTEC = Bay of Bengal Initiative for Multi-Sectoral Technical and Economic Cooperation.

Sources: UN World Tourism Organization. 2020. World Tourism Barometer and Statistical Annex, July 2021, Volume 19 and Volume 18, June 2020; and past volumes. Bangladesh 2018–2019 arrivals from Special Branch of the Bangladesh Police. Bhutan, 2015–2019 "gross earnings" Maximum Demonstrated Production Rate data provided by the Ministry of Foreign Affairs 6 July 2020. Revised figures for Thailand 2016–2019 provided by Department of Tourism. Myanmar international arrivals data are from the Ministry of Hotels and Tourism. https://tourism.gov.mm/statistics/arrivals-2019-december/. The data are presented monthly from January to October as "up to" that month, but then for November and December, the "up to" designation was not included. The total in December 2019 seems in line with previous years, that was number used.

International arrivals to BIMSTEC countries increased from 24 million in 2010 to 65 million in 2019, with arrivals from the PRC increasing from 1.4 million to 12.2 million by 2018, comprising 20% of all arrivals.[48] Thailand accounted for 10.5 million arrivals from the PRC. The second-most popular destination for visitors was Myanmar, which received nearly 1.5 million arrivals from the PRC in 2019, an increase of almost 52% over 2018.[49]

Although data are lacking on the number of visitors following circuits in BIMSTEC countries, each category of demand drivers just discussed shows the potential demand for a circuit. A recommendation of this report is to increase the capacity for market research and data collection in each country, especially for circuits.

Domestic travel will remain the most accessible market until international cross-border travel can safely resume once COVID-19 recedes (Table 15). Before the pandemic, intraregional tourism trends were moving in a favorable direction for BIMSTEC countries. The tables, which were mostly extracted from the UNWTO's *Yearbook of Tourism Statistics* and *Compendium of Tourism Statistics*, show a growing number of arrivals between several BIMSTEC countries. These trends could be important for BIMSTEC intraregional tourism development and cooperation post pandemic, especially for circuits.

The data for Thai outbound visitors require further investigation. Outbound data from the Ministry of Tourism and Sports in Thailand are substantially different for some countries than the UNWTO's outbound and country inbound data. For example, Nepal's official tourism statistics for 2018 show 53,250 arrivals from Thailand, whereas Thailand's Ministry of Tourism and Sports show 19,606. A request was pending for clarification.

Table 7 shows that data for Bangladesh intraregional arrivals is incomplete. Arrivals from India to Bangladesh dropped 27% from 105,522 in 2011 to 77,177 in 2014, but then increased to 100,176 in 2015 and continued increasing in 2019. While an explanation from tourism authorities was pending, further research is recommended.

Table 7: Intraregional Arrivals in Bangladesh

Country	2010	2011	2012	2013	2014	2015	2019
Bhutan	...	...	...	...	...	...	...
India	...	105,522	78,119	78,975	77,177	100,176	270,024
Myanmar	...	...	...	...	...	...	...
Nepal	...	...	...	...	...	1,036	2,974
Sri Lanka	...	...	...	...	...	...	...
Thailand	...	853	644	505	715	...	...

... = not available.

Note: Updated data for Thailand provided by the Department of Tourism.

Source: UN World Tourism Organization. 2021. *Yearbook of Tourism Statistics*. Madrid. p. 92.

---

[48] This estimate is derived from the UNWTO online tourism database. No outbound PRC data was reported for Bangladesh. The UNWTO Tourism Barometer reported arrivals to Bangladesh up to 2017; 2018 and 2019 arrivals in Bangladesh from Special Branch, Bangladesh Police.

[49] UNWTO. People's Republic of China Outbound Tourism: Trips Abroad by Resident Visitors to Countries of Destination, 1995–2019. February 2021 (accessed 21 August 2021 via subscription).

Table 8 shows arrivals from Bangladesh and India almost doubled from 2015 to 2018, while arrivals from Thailand dropped 68% from 2014 to 2018. According to the Tourism Council of Bhutan, a Bhutan–Thailand "Friendship Offer" initiative caused the increase in 2014. Requests were pending as of mid-December for more information to determine what drove the increases from Bangladesh and India.

Table 8: Intraregional Arrivals in Bhutan

Country	2014	2015	2016	2017	2018	2019
Bangladesh	...	5,851	8,596	10,536	10,450	13,016
India	...	91,733	138,201	172,751	191,836	230,381
Myanmar	11	18	20	139	104	216[a]
Nepal	203	192	285	265	371	1,438
Sri Lanka	27	18	28	57	63	316
Thailand	12,105	3,778	4,175	4,047	3,886	4,086

... = not available.

[a] For 2019, arrivals in Myanmar from Thailand were unavailable, but the total of overnight stays were reported in UN World Tourism Organization. 2021. *Yearbook of Tourism Statistics*. Madrid. Overnight stays in Bhutan from Myanmar increased from 482 to 2,055, and from Thailand from 16,993 to 31,650 from 2018 to 2019. The other figures for Myanmar are for arrivals of nonresidents at national borders. See *Yearbook of Tourism Statistics* data for 2015–2019.

Source: Tourism Council of Bhutan.

Arrivals from Bangladesh increased 173% from 942,562 in 2014 to 2.5 million in 2019 due to the liberalization of the Revised Travel Arrangement between India and Bangladesh in 2013 and 2018 (Table 9). The short-term medical visa was extended to 1 year and it has become easier for Bangladeshis to get multiple-entry visas for India. Student and employment visas are also easier to get. For India, making it easier for Bangladeshis to get visas proved to be a fast-track way to boost inbound tourism. This is significant for the future marketing of cross-border circuits. Further research is needed for each of the five other BIMSTEC countries to explain the increases in arrivals.

Table 9: Intraregional Arrivals in India

Country	2014	2015	2016	2017	2018	2019
Bangladesh	942,562	1,133,879	1,380,409	2,156,557	2,256,675	2,577,727
Bhutan	16,001	19,084	20,940	25,267	26,470	28,178
Myanmar	54,631	55,341	51,376	56,952	75,773	86,842
Nepal	126,416	154,720	161,097	164,018	174,096	164,040
Sri Lanka	301,601	299,513	297,418	303,590	353,684	330,861
Thailand	121,362	115,860	119,663	140,087	166,293	169,956

Source: UN World Tourism Organization. 2021. *Yearbook of Tourism Statistics*. Madrid. pp. 455, 458.

Inbound tourism to Myanmar recorded the fastest growth in South Asia and Southeast Asia as the country increasingly opened to tourism. Among BIMSTEC countries, India and Thailand were principal sources, with increases of 14.0% and 12.3%, respectively, from 2018 to 2019 (Table 10).

Table 10: Intraregional Arrivals in Myanmar

Country	2014	2015	2016	2017	2018	2019
Bangladesh	3,654	4,237	7,880	4,602	4,195	4,289
Bhutan	105	136	...	174	154	226
India	62,117	59,692	63,864	86,907	102,702	117,317
Nepal	1,511	1,496	...	2,205	2,487	3,609
Sri Lanka	3,081	3,155	...	4,318	5,050	5,729
Thailand	1,434,416	1,604,212	1,537,957	1,524,516	1,719,350	1,930,425

... = not available.

Sources: Data for Bhutan, Nepal, Sri Lanka, and Thailand from the ASEAN online tourism database (accessed 15 December 2020); UN World Tourism Organization. 2021. *Yearbook of Tourism Statistics*. Madrid. pp. 687–688.

Arrivals from India increased by 88%, Sri Lanka by 50%, and Thailand 25% from 2014 to 2019 (Table 11). According to the Nepal Ministry of Culture, Tourism and Civil Aviation, Lumbini received 204,825 visitors from India for pilgrimages in 2019, comprising 80.5% of all Indian visitors to Nepal in that year.[50] A goal of the Nepal Tourism Board's Visit Nepal Year 2020 campaign was to increase by 30% visitors from India in 2020, but this did not happen due to COVID-19.

Table 11: Intraregional Arrivals in Nepal

Country	2014	2015	2016	2017	2018	2019
Bangladesh	21,851	14,831	23,440	29,060	25,959	25,849
Bhutan	...	5,428	6,595	10,923	...	11,676
India	135,343	75,124	118,249	160,268	200,438	254,150
Myanmar	...	21,631	25,769	30,852	41,402	36,274
Sri Lanka	37,546	44,367	57,521	45,361	70,610	56,316
Thailand	33,422	30,953	26,722	39,154	53,250	41,660

... = not available.

Sources: Nepal Ministry of Culture. Tourism and Civil Aviation; World Tourism Organization. 2021. *Yearbook of Tourism Statistics*. Madrid. p. 694.

---

50   Government of Nepal, Ministry of Culture, Tourism and Civil Aviation. 2020. *Nepal Tourism Statistics 2019*. Kathmandu.

The largest intraregional BIMSTEC market for Sri Lanka was India, which increased almost 200,000 arrivals from 2014 to 2018. The other BIMSTEC markets for Sri Lanka were all less than 10,000 arrivals in 2019 (Table 12).

Table 12: Intraregional Arrivals in Sri Lanka

Country	2014	2015	2016	2017	2018	2019
Bangladesh	10,754	13,358	17,098	15,510	10,487	8,261
Bhutan	425	397	462	737	679	343
India	242,734	316,247	356,729	384,628	424,887	355,002
Myanmar	2,644	2,794	3,645	4,365	3,241	3,124
Nepal	3,319	5,801	13,153	5,144	5,302	5,414
Thailand	9,260	10,112	9,462	10,828	9,178	9,861

Sources: UN World Tourism Organization. 2021. *Yearbook of Tourism Statistics*. Madrid. pp. 893, 896, 899; Sri Lanka Tourism Development Authority. 2019. *Annual Statistical Report 2019*. Colombo. pp. 18–19.

Thailand attracted the largest number of international tourist arrivals and tourism receipts of BIMSTEC countries (Table 6). Because of this, it is worthwhile to examine arrival trends to Thailand from other BIMSTEC countries. Arrivals from India to Thailand increased 114% from 2014 to 2019 due in part to intensified marketing by the Tourism Authority of Thailand of the golf and wedding segments (Table 13). In 2013, Thailand won the Best Golf Destination award in the Safari India National Tourism Awards and Best Wedding Destination from the Hospitality India Awards. Both awards boosted tourism from India to Thailand. Inbound visitors to Thailand from India increased due to the 2015–2016 Middle East Respiratory Syndrome (MERS) outbreak in the Republic of Korea, which is believed to have diverted tourists to Thailand. A weaker baht also resulted in those Indians who would have traveled to Japan and Republic of Korea choosing Thailand instead.

Table 13: Intraregional Arrivals in Thailand

Country	2014	2015	2016	2017	2018	2019
Bangladesh	88,134	107,394	100,263	121,765	129,572	136,677
Bhutan	...	...	...	24,208	...	3,419
India	932,603	1,069,422	1,194,508	1,415,197	1,595,754	1,996,842
Myanmar	206,794	259,678	341,626	365,606	368,159	378,232
Nepal	25,887	32,678	42,486	43,251	56,000	60,377
Sri Lanka	77,441	75,434	68,195	63,267	64,760	71,043

...= not available .

Source: UN World Tourism Organization. 2021. *Yearbook of Tourism Statistics*. Madrid. p. 945, 948.

From 2016, tourism between India and Thailand was boosted by Air Asia, Nok Scoot, and Thai Smile Air (a subsidiary of Thai Airways), among others, putting on more flights. Flights also increased between Indian destinations, including Kochi, Chennai, Kolkata, Jaipur, and Bhubaneswar.[51]

---

[51]   Nok Skoot Airline was a joint venture of Thailand-based Nok Air and Singapore-based Skoot (a subsidiary of Singapore Airlines) launched in 2014. The airline closed in June 2020 due to COVID-19.

Tourism from Myanmar to Thailand rose more than 45% in 2016, due in part to a 14-day visa free scheme launched in 2015. The Myanmar outbound market, according to Mastercard,[52] has been growing steadily at a compound annual growth rate of 10.6% since 2016 before being derailed by COVID-19. Mastercard observed particularly strong growth among millennials and seniors as disposable income for these groups increased, and from their desire to see new destinations. Adventure holidays, boutique hotel stays, and solo travel helped drive this trend.

Tourism from Bangladesh to Thailand rose 55% from 2014 to 2019. The reason for this may be the formal cooperation between the countries starting in 2015 on the Buddhist circuit and wider tourism.

Tourism from Nepal to Thailand is a small market, but it more than doubled from almost 26,000 arrivals in 2014 to 60,377 in 2018. Arrivals rose 30% in 2016 from 2015 as the country recovered from the 2015 earthquake.

Tourism-Related Income and Employment Trends

GDP per capita in BIMSTEC countries has shown continual annual growth nearly every year over the past decade (Table 14). While some of this growth can probably be attributed to the growth in international and domestic tourism receipts, it is also an indicator of increased disposable income, which probably helped fuel more domestic and outbound travel. The GDP per capita figures from India, however, do not show the full picture, because India has the largest population among BIMSTEC countries. The number of outbound travelers from India more than doubled from 13.0 million in 2010 to 26.9 million in 2019, with one million visiting neighboring countries. In 2019, they spent more than $28.5 billion.[53] This trend is a positive indicator for travel and tourism in the region, since over a million Indians visited Nepal, Sri Lanka, and Thailand.[54] For the other countries, increasing income and GDP per capita have generally shown parallel growth in outbound travel.

Tourism growth would have probably continued until COVID-19 hit international travel to and between BIMSTEC countries in 2020. As the pandemic recedes with increased health protocols, testing, and vaccination among member countries—and tourism-generating markets beyond BIMSTEC—international travel and tourism is expected to recover and provide a faster-track solution for economic recovery.

Table 14: BIMSTEC Gross Domestic Product Per Capita
($)

Country	2010	2011	2012	2013	2014	2015	2016	2017	2018	2019
Bangladesh	771	820	844	992	1,104	1,224	1,373	1,510	1,638	1,808
Bhutan	2,225	2,509	2,472	2,396	2,559	2,647	2,809	3,370	3,332	3,419
India	1,408	1,534	1,505	1,531	1,611	1,671	1,763	1,998	2,089	2,151
Myanmar	...	...	...	1,208	1,275	1,192	1,136	1,139	1,205	1,276
Nepal	620	697	667	669	730	744	741	891	961	1,034
Sri Lanka	2,744	3,125	3,351	3,609	3,819	3,842	3,886	4,077	4,079	3,852
Thailand	5,174	5,590	5,956	6,259	6,029	5,903	6,058	6,624	7,329	7,843

... = not available as of 20 September 2020, BIMSTEC = Bay of Bengal Initiative for Multi-Sectoral Technical and Economic Cooperation.

Sources: Asian Development Bank. Key Indicators Database (accessed 15 December 2020); Central Bank of Sri Lanka. 2019 Annual Report.

---

52   Choong, Desmond and Wong, Dr. Yuwa Hedrick. Mastercard Future of Outbound Travel in Asia Pacific (2016-2021) Report: https://newsroom.mastercard.com/asia-pacific/files/2017/01/Mastercard-Future-of-Outbound-Travel-Report-2016-2021-Asia-Pacific1.pdf, p. 2
53   World Tourism Organization. 2020 Compendium of Tourism Statistics dataset [Electronic]. Madrid. (Data updated 23 November 2020).
54   UNWTO. Data on Outbound Tourism (2020); Government of India, Ministry of Tourism. India Tourism Statistics at a Glance 2019. Madrid.

Domestic Tourism: Overnight Visitors

India and Thailand saw substantial increases in domestic overnight visitors from 2014 to 2018 (Table 15).[55] Domestic markets are a high priority for most BIMSTEC countries since international travel will likely still be highly restricted into 2022 because COVID-19.

Table 15: BIMSTEC Domestic Overnight Tourist Visits
('000)

Country	2014	2015	2016	2017	2018	2019
Bangladesh	...	...	...	...	10	15
Bhutan	...	...	...	...	...	765
India	1,283,000	1,432,974	1,615,389	1,657,546	1,853,788	2,322,983
Myanmar	...	...	...	...	...	...
Nepal	...	...	...	...	...	...
Sri Lanka	2,585	2,225	2,539	2,725	2,950	2,280
Thailand	98,396	106,841	115,568	125,471	130,868	131,559

... = not available, BIMSTEC = Bay of Bengal Initiative for Multi-Sectoral Technical and Economic Cooperation.

Sources: UN World Tourism Organization. Compendium of Tourism Statistics dataset [electronic] (accessed 21 August 2021); Bangladesh 2018–2019 Tourism Stakeholders' Opinion.

The WTTC, through its online data tool, reported details of the growth in domestic tourism spending for all BIMSTEC countries, except Bhutan. Domestic tourism spending in India grew steadily from 2013 to 2019, totaling $143 billion by 2019 (Figure 9). For the other BIMSTEC countries combined, it totaled $41 billion in 2019, up from almost $24 billion in 2010 (Table 16). Most of this spending was concentrated in Thailand, where it increased from $16.3 billion in 2010 to $28.5 billion by 2019, accounting for 69% of BIMSTEC's non-India domestic tourism spending.

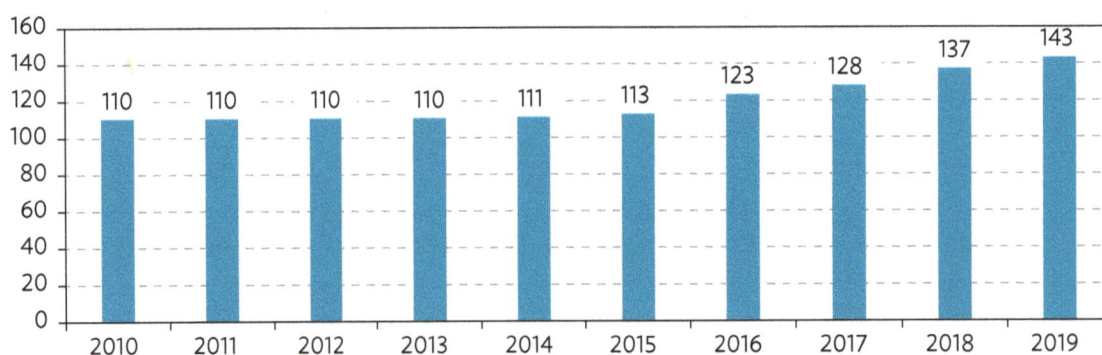

Figure 9: Domestic Tourism Spending in India, 2010–2019
($ billion)

Source: World Travel & Tourism Council. Online Data Tool. http://tool.wttc.org (accessed 11 December 2020).

---

55   UNWTO. 2019. *Methodological Notes to the Tourism Statistics Database.* Madrid. p. 24.

Table 16: Domestic Tourism Spending in BIMSTEC Countries
($ billion)

Country	2010	2011	2012	2013	2014	2015	2016	2017	2018	2019
Bangladesh	4.6	4.6	4.9	5.1	5.7	6.0	6.1	6.5	7.3	7.8
Bhutan	...	...	...	...	...	...	...	...	...	0.1
India	110	110	110	110	111	113	123	128	137	143
Myanmar	0.7	0.6	0.9	0.8	0.8	0.9	1.1	1.2	1.1	1.2
Nepal	0.8	0.9	0.8	0.8	0.9	0.9	1.0	0.9	1.0	1.0
Sri Lanka	1.5	1.9	1.9	1.8	2.0	2.2	2.4	2.4	2.5	2.5
Thailand	16.3	18.5	21.3	24.2	26.8	24.9	25.7	26.2	27.5	28.5

... = not available.

Note: Table 16 lists members of the Bay of Bengal Initiative for Multi-Sectoral Technical and Economic Cooperation (BIMSTEC)

Source: World Travel & Tourism Council. Online Data Tool. http://tool.wttc.org (accessed 11 December 2020).

Impact of COVID-19 on Tourism in BIMSTEC Countries

As of 21 August 2021, the CDC was recommending against all travel to 74 economies, including BIMSTEC members of Bangladesh, Nepal, and Thailand. Sri Lanka was at Level 3 (high risk, recommended travel only if vaccinated), India was at a Level 2 (moderate risk), and Bhutan was on Level 1.[56] It is important to note that these warning levels were changing every week in 2021 depending on levels of COVID-19 spread and risk. As Table 5 shows, much of the world was still risky for travel as of late August 2021.

The July 2021 issue of the UNWTO World Tourism Barometer presented limited but impactful data on percentage changes in international arrivals and receipts for the first 5 months of 2021, compared with 2020 and 2019.

The drop in arrivals and receipts caused by the COVID-19 severely affected travel- and tourism-related employment and GDP in BIMSTEC states (Table 17 and Table 18). Table 19 shows that tourism before the pandemic was a substantial source of employment and GDP for BIMSTEC countries and shows the losses each BIMSTEC state suffered in 2020. Most BIMSTEC states experienced dramatic drops in tourism receipts and employment and the sector's contribution to GDP, including India (8.3 million jobs, $69 billion); Myanmar (400,000 jobs, $3.5 billion, more than half the usual contribution to GDP); Nepal ($900 million, about half the normal contribution); Sri Lanka (214,000 jobs and $4.9 billion, more than half); and Thailand ($64.8 billion, more than half). BIMSTEC states lost an estimated $177.4 billion in GDP and 42.45 million tourism-related jobs in 2020 due to the economic impact of COVID-19.

---

[56]   Centers for Disease Control and Prevention, COVID-19 Travel Recommendations by Destination (accessed 21 August 2021).

Table 17: Percentage Declines in International Tourist Arrivals to BIMSTEC Countries

Country	2021 vs. 2020						2021 vs. 2019					
	YTD	Jan	Feb	Mar	Apr	May	YTD	Jan	Feb	Mar	Apr	May
Bangladesh	…	…	…	…	…	…	…	…	…	…	…	…
Bhutan	(100)	(100)	(100)	(100)	…	…	(100)	…	…	…	(100)	…
India	(84.6)	(91.4)	(91.4)	(59.8)	…	…	(90.5)	(91.3)	(92.0)	(87.4)	(91.0)	…
Myanmar	…	…	…	…	…	…	…	…	…	…	…	…
Nepal	(74.2)	(88.9)	(90.7)	(65.0)	…	…	(88.6)	(89.1)	(91.1)	(88.2)	(79.5)	(98.1)
Sri Lanka	(97.0)	(99.3)	(98.4)	(93.6)	…	…	(98.4)	(99.3)	(98.7)	(98.1)	(97.5)	(96.0)
Thailand	(99.5)	(99.8)	(99.7)	(99.2)	…	…	(99.8)	(99.8)	(99.8)	(99.8)	(99.7)	(99.8)

… = not available, ( ) = negative, YTD = year to date.

Note: Data for Myanmar 21/20 and 21/19 was not reported to UNWTO.

Source: UN World Tourism Organization. 2021. *World Tourism Barometer and Statistical Annex*. July 2021, Volume 19.

Table 18: Percentage Declines in International Tourist Receipts for BIMSTEC Countries

Country	2021 vs. 2020						2021 vs. 2019					
	YTD	Jan	Feb	Mar	Apr	May	YTD	Jan	Feb	Mar	Apr	May
Bangladesh	(72.7)	…	…	…	…	…	(75.1)	…	…	…	…	…
Bhutan	(100)	(100)	(100)	(100)	…	…	(100)	…	…	…	(100)	…
India	…	…	…	…	…	…	…	…	…	…	…	…
Myanmar	…	…	…	…	…	…	…	…	…	…	…	…
Nepal	(80.7)	(91.8)	(89.1)	(54.6)	(63.5)	(56.8)	(89.9)	(90.2)	(89.9)	(84.4)	(92.6)	(92.0)
Sri Lanka	(97.3)	(99.3)	(98.4)	(93.6)	…	…	(98.9)	(99.5)	(99.0)	(98.7)	(98.2)	…
Thailand	(90.9)	…	…	…	…	…	(94.1)	…	…	…	…	…

… = not available, ( ) = negative, BIMSTEC = Bay of Bengal Initiative for Multi-Sectoral Technical and Economic Cooperation, YTD = year to date.

Source: UN World Tourism Organization. 2021. *World Tourism Barometer and Statistical Annex*, July 2021, Volume 19.

Table 19: Contribution of Tourism to Total Employment and GDP in BIMSTEC Countries, 2019 and 2020

Country	Total Employment (%) 2019/2020	Total Employment 2019/2020	Total Contribution to GDP (%) 2019/2020	Total GDP ($) 2019/2020
Bangladesh	2.9/2.3	1.86/1.45 million	2.7/1.7	$9.4/$6.3 billion
Bhutan	6	50,000+	7 (est.)	$345 million
India	8.8/7.3	40/31.7 million	6.9/4.7	$191/$122 billion
Myanmar	6.3/4.8	1.4/1 million	5.9/2.2	$5.5/$2 billion
Nepal	6.9/5.5	1 million/835,000	6.7/3.6	$2/$1.1 billion
Sri Lanka	10.9/8.4	888,000/674,000	10.4/4.9	$8.9/$4 billion
Thailand	21.4/18.4	8/6.8 million	20/8.4	$106.5/$41.7 billion

BIMSTEC = Bay of Bengal Initiative for Multi-Sectoral Technical and Economic Cooperation, GDP = gross domestic product.

Note: World Travel & Tourism Council data were unavailable.

Source: WTTC Economic Impact Reports; Tourism Council of Bhutan.

Key Issues Affecting BIMSTEC Tourism

For the tourism industry in the BIMSTEC region to recover from COVID-19 and return to its resilient upward trajectory before the pandemic, the following need to be considered to help build thematic circuits for regional and national tourism: pandemic recovery time, infrastructure, marketing and product, human resources, policy, and investment. This section reviews each issue area, using the country rankings of the World Economic Forum's 2019 Tourism Competitiveness Index (WEF-TTCI).[57] The rankings help identify certain areas of need in each country that affect the potential for growing regional tourism and cross-border thematic circuits. Solutions for tackling these needs form the revised action plan based on strategic objectives and recommendations presented in this report.

Issue Area #1: COVID-19 Recovery

Controlling the spread of and recovering from COVID-19 through vaccination programs and widely communicated health and safety protocols is clearly the top priority for every country. Tourism everywhere will not be able to return to normal unless uniform health and safety protocols, controls, and protection are effectively communicated and visibly implemented across the entire tourism value chain. When it is safe to travel again, visitor confidence can be restored, and tourism can get back on track to becoming a potent force for economic and social development, as well as cultural and natural heritage protection. Pursuing the national and cross-border circuits could be an optimal means for expediting a return to pre-COVID-19 tourism growth among BIMSTEC countries because they could help advance cooperation on the joint cross-border implementation of health and safety protocols.

Issue Area #2: Recovering Lost Tourism Jobs and Businesses

An estimated 42 million tourism-related jobs were lost in BIMSTEC countries due to COVID-19. This figure included direct and indirect jobs across the tourism value chains in each BIMSTEC country.

At ADB's 25 November 2020 workshop, where a summary of this report was presented, participants were asked to estimate how many tour operators and hotels had to close and how many people lost tourism-related jobs in their countries due to COVID-19. Table 20 presents their unofficial estimates. There are probably underestimates considering that apart from some limited domestic tourism, the industry closed completely in most BIMSTEC countries, as well as in the generating country markets outside the BIMSTEC region.

Table 20: COVID-19 Impacts on BIMSTEC Tourism

Country	Estimated Number of Tour Operators that Closed	Estimated Number of Hotels that Closed	Number of Tourism-Related Jobs Lost
Bangladesh	250 to 1,200+	150 to 300+	25,000 to 100,000
Bhutan	All	150 or 96%–99% of all hotels	49,000 or 98% of all tourism jobs; few displaced are being employed in other work
India	No response	No response	No response
Myanmar	90%–100% of total	60% of total or above	80% of all tourism jobs
Nepal	3,000+	95%	30,000 direct and more than 200,000 indirect jobs were lost
Sri Lanka	4,000	5,000	1 million
Thailand	Data not available	Data not available	60%–70% of tourism businesses closed during the pandemic due, in part, to a government-imposed national lockdown policy.

BIMSTEC = Bay of Bengal Initiative for Multi-Sectoral Technical and Economic Cooperation.
Source: Asian Development Bank. Workshop Survey. 25 November 2020.

---

[57] Bhutan was included in the 2017 WEF-TTCI and Myanmar in 2015, but neither were in the 2019 survey due to insufficient data.

In Thailand, some hotels and tour operators were temporarily closed during the pandemic. Precise data on this are unavailable, which underscores the priority need to assess the damage that COVID-19 has caused to the industry.

Issue Area #3: Tourism Infrastructure Issues and Needs

Infrastructure improvements in BIMSTEC countries will be essential for further leveraging the region's tourism potential and cross-border thematic circuits. All member countries recognize this need and are pursuing national infrastructure and tourism plans to enhance existing tourism offers and produce new ones.

The following identifies some of the infrastructure needs of BIMSTEC countries:

- In Bangladesh, road quality and port improvements have been highlighted by the government. Progress is being made in visitor facilities at multiple sites, several of which are on circuit routes.
- The Government of India announced on 15 August 2021 that it will invest $1.35 trillion in infrastructure, especially in logistics, to improve connectivity and spur economic development.[58]
- Myanmar's Tourism Master Plan cited a need for better transport infrastructure for, among other things, increasing access to electricity, telecommunications, health, and hygiene services.
- Nepal's vision recognizes that air connectivity and the national carrier need to be further improved.
- The Government of Bhutan's 2016 Review Report recognizes that the benefits of tourism are not equally distributed throughout the country, particularly in the eastern region, due to poor connectivity.[59]
- Thailand's Second National Tourism Development Plan cites the importance of regional connectivity and corridors, which would boost multi-country thematic circuits, but greater progress on implementation is needed.[60]

The tourism sector industry's recovery can be achieved partly through cross-border thematic circuits, such as the Buddhist circuit, but this will depend on especially the following types of infrastructure:

- transport connectivity in BIMSTEC countries;
- tourist service infrastructure (as defined by the WEF-TTCI);
- additional tourist infrastructure on the services side, such as restaurants and hotels, visitor amenities, and site management facilities;
- health, sanitation, and waste management infrastructure; and
- information and communication technology infrastructure (WiFi, broadband, mobile connectivity).

The Secretariat has an important role in helping BIMSTEC countries improve each area of infrastructure, especially in sharing best practices and harmonizing standards and regulations across borders. The latter is especially important for facilitating the development and promotion of marketable cross-border thematic circuits. BIMSTEC countries reaffirmed their commitment to this at the Fourth BIMSTEC Summit in Kathmandu, held in August 2018. In the summit declaration, members agreed to build on the 2005 Plan of Action for Tourism Development and Promotion for the BIMSTEC Region for Tourism Development and Promotion by focusing on the development of circuits, including the Buddhist and Temple Tourist circuits and other product segments. For private sector operators, better infrastructure makes these circuits easier to visit and thus market.

---

58   A. Chaudhary and S. Ghosh. 2021. India Aims to Spend $1.4 Trillion Building Infrastructure. Bloomberg. 15 August.
59   Government of Bhutan. *Review Report on Tourism Policy and Strategies of the Economic Affairs Committee.* 2016.
60   Responses sent mid-August 2020 by the Ministry of Tourism on the implementation status of the plan's five strategies.

As Table 22 shows, WEF-TTCI rankings also highlight the need to improve the tourism infrastructure of BIMSTEC countries, including the number and quality of hotel rooms, resorts, entertainment centers, and car rental companies. The WEF-TCCI does not cover all aspects of tourism infrastructure, and some areas that are missing are perhaps more important for setting up and maintaining thematic circuits and the sites along circuits. These include health and hygiene facilities, access to visitor information, waste management, visitor flow at sites, internet access, and signage.

Some of the priority BIMSTEC circuits for infrastructure improvements could include the following:

- buddhist circuit for Nepal and India;
- himalaya circuit between Nepal and Bhutan;
- river cruise circuit between Bangladesh and India;
- ocean cruise circuit between Sri Lanka and Kerala, India; and
- heritage site circuit from Thailand to Myanmar, starting from Viet Nam.

These are all circuits that regional tour operators were marketing before the outbreak of the COVID-19 pandemic. Market data are unavailable on the level of demand. For most of 2020, these circuits saw little, if any, activity. An assessment of regional and national tour operators that are still in business will be needed to determine who can still do tours to these circuits, and what is needed to stimulate this business, which will also help to attract new operators.

The Swadesh Darshan thematic circuit development program in India provides a useful toolkit and a list of infrastructure needs that are being addressed that could help guide the development of thematic circuits across the BIMSTEC region. These needs are described in detail in the India section of this report and are summarized in Figure 17. These infrastructure needs are supported financially by the program. Annex 2 has a synopsis of the toolkit.

With COVID-19 severely affecting travel and tourism globally, health and hygiene have proven to be critical infrastructure elements, without which, as 2020 has shown, entire economies have been paralyzed. All countries raced to impose measures and protocols to control the spread of COVID-19. Their measures are summarized in the succeeding country profiles. These restrictions have resulted, as mentioned before, in widespread closures and bankruptcies of travel-related businesses, including attractions, restaurants and hotels. Indeed, COVID-19 has compelled many countries to take extraordinary health and hygiene measures, which might result in higher WEF-TTCI rankings in the next survey (Table 21).

Table 21: World Economic Forum's Travel and Tourism Competitiveness Index Health Rankings in BIMSTEC Countries

Country	Health and Hygiene	Physician Density	Use of Basic Sanitation	Use of Basic Drinking Water
Bangladesh	103	103	112	64
Bhutan	96	107	108	1
India	105	100	115	...
Myanmar	115	136	134	124
Nepal	93	137	126	...
Sri Lanka	69	82	79	54
Thailand	88	97	58	55

... = not available, BIMSTEC = Bay of Bengal Initiative for Multi-Sectoral Technical and Economic Cooperation, WEF-TTCI = World Economic Forum's Travel and Tourism Competitiveness Index.

The 2019 survey included all BIMSTEC countries except Bhutan, which was included in the 2017 survey and Myanmar, which was included in the 2015 survey.

Source: WEF-TTCI.

Issue Area #4: Transport Infrastructure Connectivity

BIMSTEC states recognize the critical importance of transport connectivity for developing domestic and cross-border tourism in the region. A master plan for BIMSTEC transport connectivity was announced in 2018 and was expected to enhance multimodal transport connectivity and help restore connectivity lost because of the COVID-19 pandemic. The plan reportedly includes a detailed review of transport issues and recommended policies and strategies for road, rail maritime, and inland water transport, as well for aviation and airports.

Pre-COVID-19 transport connectivity improvements in BIMSTEC states, including more low-cost carrier routes and airlines, new and improved airports, improved roads, and ports, helped boost tourism in the BIMSTEC region from 24 million visitors in 2010 to almost 66 million in 2019. But as the WEF-TTCI 2019 rankings for air, ground, and port infrastructure show there is room for improvement among most member countries in each of these areas (Table 22).

Table 22: World Economic Forum's Travel and Tourism Competitiveness Index 2019 Rankings of BIMSTEC countries

Country (2019 Index ranked 140 countries)	Air Transport Infrastructure (overall access)	Quality of Air Transport Infrastructure (airports)	Quality of Road Infrastructure	Quality of Port Infrastructure	Tourist Services Infrastructure (availability of services)	Quality of Tourism Infrastructure (hotels and resorts)
Bangladesh	111	115	111	89	133	116
Bhutan (2017/136) See note below	67	103	79	...	109	62
India	33	57	51	51	109	132
Myanmar (2015/141) See note below	115	136	134	124	137	124
Nepal	93	137	126	...	126	80
Sri Lanka	69	82	79	54	92	35
Thailand	22	42	55	66	14	15

... = not available, BIMSTEC = Bay of Bengal Initiative for Multi-Sectoral Technical and Economic Cooperation, WEF-TTCI = World Economic Forum's Travel and Tourism Competitiveness Index.

Note: Bhutan was last ranked in 2017 and Myanmar in 2015.

Source: WEF-TTCI.

The WEF-TTCI rankings are useful in highlighting areas of transport infrastructure needs and gaps, especially where countries have low rankings. Closing these gaps will facilitate travel within and between BIMSTEC countries, as well as with countries outside the region. Some of the gaps, which resulted in lower WEF-TTCI rankings, will probably be tackled by the many project initiatives of the BIMSTEC Master Plan for Transport Connectivity and the BIMSTEC Motor Vehicle Agreement.

The master plan was approved at the 17th BIMSTEC Ministerial Meeting and adopted at the Fifth BIMSTEC Summit. According to a joint 2018 ADB and BIMSTEC report, at least 10 border roads (1 in Bangladesh, 2 in India, 4 in Myanmar, 2 in Nepal, and 1 in Thailand) are expected to be upgraded under the transport master plan, which will also help the Trilateral Highway linking India, Myanmar, and Thailand.[61] The plan has projects to improve roads in West Bengal and Bihar; road-based circuits in India, Nepal, Bhutan, and Sri Lanka that connect historical Buddhist sites; airports in all BIMSTEC countries; and border crossings between Nepal and India; and Bangladesh and India. All these projects should facilitate cross-border and intra- and inbound travel to BIMSTEC countries, which will enhance tourism not only for Buddhist circuits but also for other circuits between countries, as well as for overall tourism attractions. Temple Tiger Group and other tourism firms will benefit from these enhancements.

The COVID-19 pandemic hit aviation hard in BIMSTEC countries and throughout Asia and the Pacific, grounding the sector. Before the pandemic, increased aviation connectivity, especially among low-cost carriers, spurred tourism into and among BIMSTEC countries. These included flights with the low-cost carriers AirAsia Group, Bangkok Airways, NokAir, NokScoot, Japan Airlines, and Thai Airways, all of which either went bankrupt, reorganized, or reduced services due to COVID-19. In 2021, this list seemed to be changing every week with many airlines having to go bankrupt or otherwise reorganizing their financing. Aviation connectivity should be continually reassessed since it is a factor in determining generating markets for BIMSTEC and will affect the tourist potential of circuits.

Issue Area #5: Marketing and Product

All BIMSTEC countries are rich in cultural and nature-based tourism assets, attractions, and activities, many of which lend themselves to forming part of two or more country circuits or itineraries. These assets have attracted visitors—either independently or in groups—in varying numbers over many years and provide a basis for expansion. The wellness segment, which includes medical tourism, was a fast-growing segment, particularly for India, Sri Lanka, and Thailand—and it will be an increasingly important post-pandemic segment, which could be linked to multiple circuits or perhaps even become a separate circuit.

Other circuit-based product offers include the Buddhist trail across India and Nepal; the Swadesh Darshan circuits of India; the Himalaya circuit between Nepal and Bhutan; the River Cruise circuit between Bangladesh and India; the Ocean Cruise circuit between Sri Lanka and Kerala, India; and the Heritage Site circuit from Thailand to Myanmar, starting from Viet Nam. However, coordination among national tourism administrations and with the private sector is lacking across the BIMSTEC region.

Without this coordination and joint branding, market research, and marketing by member governments, BIMSTEC as a region will lack coordinated regional initiatives that are visible in the travel and tourism sector. BIMSTEC states are targeting the same markets, but there is little cooperation in marketing to and collecting data about these target markets. Cost-sharing among BIMSTEC states in launching familiarization trips involving media, social media influencers, and tour operators has not yet started.

Market research needs and the quality of data on the travel and tourism market vary from country to country in the BIMSTEC region, but at the regional level, coordinated data collection similar to what ASEAN is doing is lacking, putting the region at a competitive disadvantage.

---

[61]   ADB. 2018. *Updating and Enhancement of the BIMSTEC Transport Infrastructure and Logistics Study.* Manila.

BIMSTEC has no tourism marketing strategy that can identify common targets based on joint research, implement the strategy in coordination with member countries, and facilitate cross-border thematic circuits. A tourism market strategy is needed. Digital marketing, which is done by ASEAN and GMS countries, should be an integral part of such a strategy, especially given the rise in digital media access in 2020 and 2021, a likely growing trend over the next several years.

Online information on the BIMSTEC region as tourist destination is lacking. There is no regional tourist information website, like the ASEAN's which brings together travel information on activities, sights, events, and tourism packages, as well as suggested itineraries for all 10 member countries (Figure 10).

Figure 10: Association of Southeast Asian Nations Tourism Website

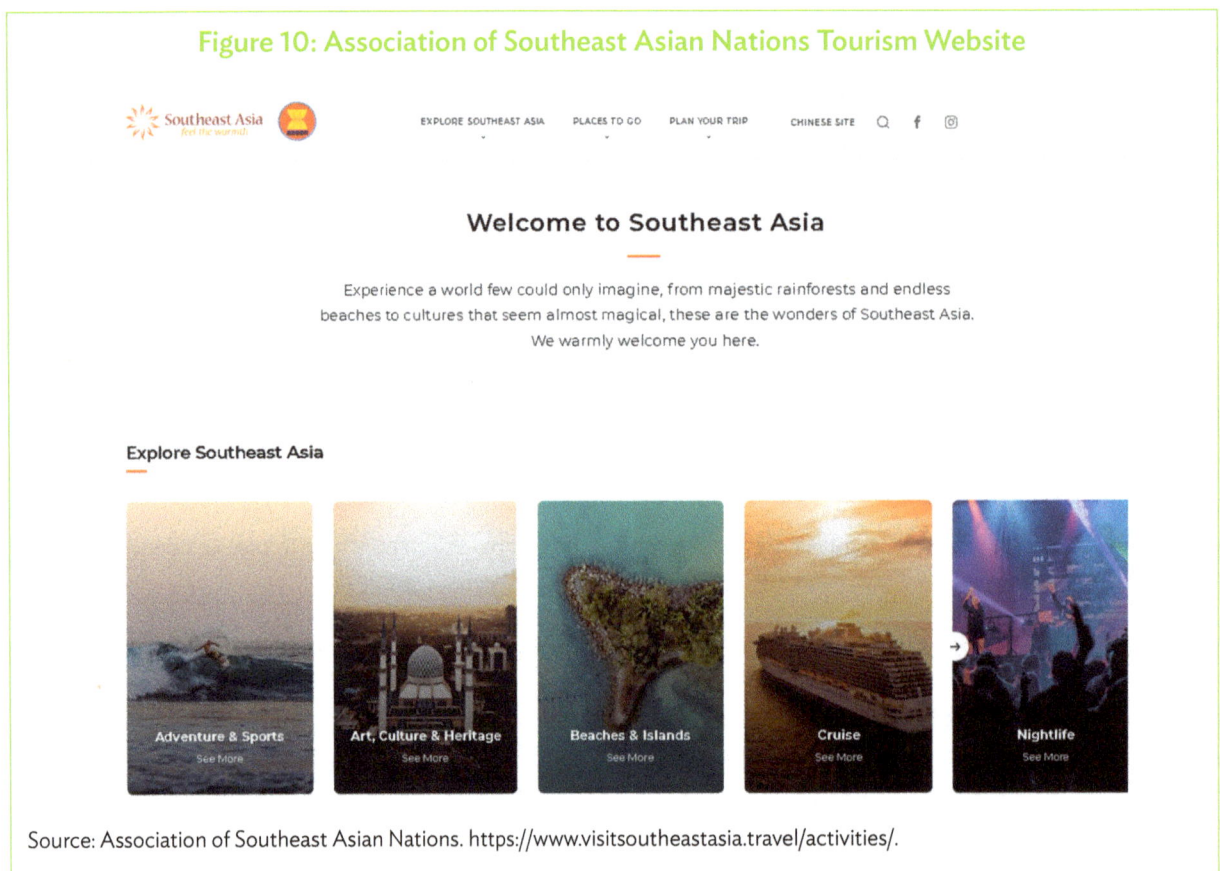

Source: Association of Southeast Asian Nations. https://www.visitsoutheastasia.travel/activities/.

Mekong Tourism has a regional marketing website like ASEAN's. This gives visitor and trade information for six GMS countries, and has a section on multi-country journeys, such as a four-country river cruise and an east–west corridor journey through the Lao PDR, Thailand, and Viet Nam.

No BIMSTEC-wide product quality standards for hotels exist. Each country has its own hotel classification system, based on star ratings. It is not clear whether a five-star hotel that is not part of an international chain, such as Marriott International or Taj Hotels, has the same standard in each BIMSTEC country. Platforms such as TripAdvisor and Booking.com do rate hotels, but these are done by user reviews and may not cover all BIMSTEC countries. It would be helpful for a prospective visitor or tour operator planning to visit a circuit to know that a five-star hotel in one BIMSTEC country would be of the same quality level in the others. This will be increasingly important as countries emerge from the COVID-19 pandemic and need to reassure visitors and operators of the quality of hotels, particularly that high standards are maintained for health and safety across all BIMSTEC countries.

From a marketing and product perspective, each BIMSTEC country needs to get an accurate picture of the businesses in the travel and tourism market have survived the COVID-19 pandemic and the number of workers in the sector that lost their jobs. In each BIMSTEC country, there have been hundreds, if not thousands, of bankruptcies affecting all types of tourism-related businesses. This exercise should also be done for all generating markets for BIMSTEC countries. It should be noted, however, that domestic tourism in some BIMSTEC countries, including India and Nepal, has increased, which has helped the reemployment of some of those who either lost their jobs or had been furloughed.

The following are specific marketing and product challenges of member countries:

- **Lack of market data and research.** Bangladesh reported annual tourism data to UNWTO up to 2019, but not monthly data, which probably contributed to its low WEF-TTCI ranking (128th out of 140 countries in the 2019 survey) for marketing and branding. Updated market data would enable all BIMSTEC countries to market more effectively, especially for their domestic circuits and for cross-border thematic circuits, such as India's Swadesh Darshan Scheme.
- **Insufficient resources.** Nepal, which was attracting growing numbers of visitors pre-COVID-19, noted in its national tourism plan—Tourism Vision 2020—that its public resources are insufficient for "massive publicity and consumer promotion." This may account for its low WEF-TTC ranking (128th on resources). The COVID-19 pandemic stretched these resources further. Joint marketing via BIMSTEC could help Nepal and other resource-constrained countries in the region.
- **Lack of a holistic approach.** Sri Lanka, in its national tourism strategy, recognizes the lack of a holistic marketing approach has affected its marketing and promotion efforts—and this may account for its relatively low WEF-TTCI ranking (97th). The country's tourism strategy also recognizes the absence of quality-driven, professional, and digitally savvy strategic activity plans.

Issue Area #6: Human Resources

The human toll of the pandemic in terms of lost jobs has been worse than ever, with millions of jobs lost in BIMSTEC countries. As travel and tourism returns, surviving businesses will try to reemploy staff—and demand surges are expected. But it will be important to fill positions with trained and educated people.

Before COVID-19, according to the WEF-TTCI travel and tourism businesses in some BIMSTEC countries faced a shortage of qualified workers. As Table 23 shows, some countries would benefit from support for building their tourism workforce. India and Sri Lanka seem better positioned in finding skilled labor.

WEF-TTCI health rankings show most BIMSTEC countries need to improve health and hygiene measures for travel and tourism (Table 21). COVID-19 has prompted all BIMSTEC countries to improve on these measures, especially focusing on human resources, thus the need to expand health and safety training and education of tourism staff.

Expanding tour guide training, particularly for visitors on multi-destination circuits, is an important human-resources aspect that needs tackling. Most, if not all, tours are conducted by private operators and guides, who are either current staff or hired by operators. No standard government-issued license for tour guides recognized across BIMSTEC states exists. Having one would allow operators to hire the same guide across borders.

Table 23: World Economic Forum's Travel and Tourism Competitiveness Human Resources 2019 Rankings in BIMSTEC Countries

Country Rankings (2019 Index ranked 140 countries)	Human Resources and Labor Market	Qualification of the Labor Market	Labor Market Availability	Ease of Finding Skilled Workers
Bangladesh	120	107	127	105
Bhutan (2017/136) See note below	89	...	...	114
India	76	78	93	34
Myanmar (2015/141) See note below	117	121	63	139
Nepal	83	98	67	95
Sri Lanka	84	49	130	57
Thailand	27	26	44	88

... = data not available, BIMSTEC = Bay of Bengal Initiative for Multi-Sectoral Technical and Economic Cooperation.

Note: Bhutan was last ranked in 2017 and Myanmar in 2015.

Source: World Economic Forum's Travel and Tourism Competitiveness Index 2019

Issue Area #7: Policy, Governance, and Investment

Every BIMSTEC country has a tourism ministry—except Bhutan, which has a national tourism council. All member countries have national tourism policies and strategies, several of which identify thematic circuits as an important area for development. While cross-border thematic circuits were highlighted as action targets in the BIMSTEC 2006 Tourism Action Plan and discussed in subsequent meetings, regional coordination on tourism policy and governance has been lacking.

Regional tourism initiatives, such as cross-border circuits, require travel across borders. Visas are often required for this unless obviated by bilateral agreements. A tourist who wants to visit more than one BIMSTEC country might have to apply to multiple embassies to get the necessary visas, which can be a disincentive for tourists and tour operators. Where agreements on visa policies among BIMSTEC members are in place, tourism in participating countries has benefited. For example, tourism from Myanmar to Thailand rose 45% after a 14-day visa free was launched. Similarly, tourism from Bangladesh to India increased 173% from 2014 to 2019 after India liberalized its Revised Travel Arrangement, which made it easier for Bangladeshis to get multiple-entry visas (Table 9). For visa liberalization to benefit all BIMSTEC member countries, each country would have to negotiate an agreement with every other member, amounting to 42 separate agreements (six per member country).

A coordinated BIMSTEC approach to facilitating seamless travel within the region is lacking, but such an approach would benefit tourism in all member countries, especially for cross-border circuits. The solution might be to either have a single visa for travel in the region or visa-free travel for certain nationalities.

A coordinated BIMSTEC approach could help create a single aviation market in the region through a regional Open Skies agreement. Over the past decade, Asia has been the fastest-growing region for air connectivity, according to IATA in a November 2020 report, which says that most of the world's top aviation markets are set to be in Asia and the Pacific, with India and the PRC driving much of this growth. India and Thailand were the third and fourth most internationally connected countries in Asia and the Pacific before COVID-19. Air connectivity within and among BIMSTEC countries was severely disrupted by the pandemic, as was the rest of Asia and the Pacific. "Restoring air connectivity in Asia is essential to support the recovery of its economies reliant on trade and linked into global supply chains ... and vital for supporting tourism flows," says IATA.[62]

Establishing a common aviation market among BIMSTEC countries would boost tourism, especially for cross-border circuits. It could also intensify cooperation in improving the quality of air transport infrastructure, particularly for countries ranking low in the WEF-TTCI in this indicator (Table 22).

The Secretariat has limited institutional resources and capacity to serve in a coordinating role to implement activities to enhance governance. The BIMSTEC Tourism Working Group, which was set up under the 2006 Tourism Action Plan, was intended to provide decision-making capacity on the plan, but it has been impeded by the need for a full consensus of members for any activity, including through virtual meetings. Meetings and workshops can also take some time to get a full consensus of members. This slows down the process of jointly managing and marketing of circuits. Even so, tour operators continued with taking visitors through a circuit, given the market potential.

For investments in cross-border circuits, since more than one country would be involved, accessing the necessary regulations and assembling the required documentation from multiple countries can be time-consuming and a disincentive for investment. Accessing all information for this, including investment incentives and prospectuses on opportunities from all BIMSTEC members, through a single portal could attract more interest in investments in BIMSTEC cross-border circuits. A tourism investment strategy, especially one that connects circuit opportunities, would be beneficial. For example, a potential investor in wayside amenities along cross-border circuits would be more likely to invest in multiple countries if the regulations were the same.

---

62   International Air Transport Association. 2020. *Air Connectivity, Measuring the Connections that Drive Economic Growth.* Montreal. p. 57.

BIMSTEC Tourism Action Plan: Addressing Issues and Realizing Circuit Tourism Opportunities

BIMSTEC meetings were conducted in February 2005 and August 2006 (ministerial meetings), September 2013 (working group meeting), and in 2018 (Fourth BIMSTEC Summit). At these meetings, BIMSTEC countries recognized that they would benefit from closer regional cooperation for addressing issues of common concern and realizing opportunities, such as cross-border circuit tourism to leverage the region's rich tourism-related assets. Officials vowed to increase tourism by fostering closer cooperation in transport networks, creating intraregional thematic tour packages, simplifying cross-border immigration procedures, sharing best practices, and providing a "clear, output-based plan of action." The latter included a focus on developing and promoting circuits, such as the Buddhist circuit, Temple Tourist circuit, and other product segments. The Plan of Action for Tourism Development and Promotion for the BIMSTEC Region was reaffirmed and the potential vision statement for BIMSTEC tourism described below was informally approved by member country representatives in an online survey at an ADB BIMSTEC tourism workshop on 25 November 2020. A draft mission statement has been added since the workshop.

In August 2006, a second tourism ministerial meeting was held in Kathmandu at which the Plan of Action for Tourism Development and Promotion for the BIMSTEC Region was adopted and the BIMSTEC Working Group on Tourism and a BIMSTEC Tourism Fund set up. The working group did not meet again until 23 September 2013, when it reaffirmed the call for intraregional cooperation and agreed on contributions of $10,000 from each member country.

The BIMSTEC Network of Tour Operators held its first meeting in 2017 and a second meeting on 8–9 December 2020. The Plan of Action for Tourism Development and Promotion for the BIMSTEC Region had not been updated since it was agreed on in 2006. In the absence of an updated plan and more recent working group meetings, not much progress has been achieved on the plan, especially progress on the goal of a regional and integrated approach to circuit development.

Because the Secretariat's resources are limited, updating the Plan of Action for Tourism Development and Promotion for the BIMSTEC Region to build on what has already been agreed upon by member countries would be an efficient way forward. The proposed updated plan described herewith builds on the original recommendations agreed in the 2005 Kolkata and 2006 Kathmandu declarations on the plan. In an online survey done at the BIMSTEC tourism workshop, which presented an earlier version of this report, representatives from all BIMSTEC member governments affirmed that the recommendations are still valid and needed. These recommendations are integrated into the plan proposed by this report.

Updating the plan will help address the issue areas discussed earlier and advance thematic circuits and, indeed, tourism overall in BIMSTEC states. In addressing these issues, it is useful to review first the features of the BIMSTEC Tourism Action Plan, followed by sample circuit itineraries, a review of other Asian region plans and strategies, and a summary of tourism profiles for each BIMSTEC member country. Highlights and best practices, where relevant, will be factored into the proposed updated plan, including a vision statement, strategic objectives, action areas, and recommendations to address issues on a short-, medium- and longer-term basis.

The Plan of Action for Tourism Development and Promotion for the BIMSTEC Region

Table 24: 2006 Plan of Action for Tourism Development and Promotion for the BIMSTEC Region

Action Title	Proposed Action	Status
BIMSTEC Information Center	BIMSTEC Tourism Information Center to be set up in India to produce and update publicity and collateral material, including brochures and CDs based on the product information provided by each member country. Compile a compendium of hotel facilities and recognized tour operators. All member countries to designate their focal points and contact information. The center will specify the format for countries to submit their information by December 2006.	Although the center has not yet been established, the Second Meeting of the BIMSTEC Network of Tour Operators held virtually 8–9 December 2020 in Colombo, Sri Lanka recommended that a virtual center be set up in India. India will design the online repository for sharing tourism information of each member state. Each member can then upload their information via a password-protected site.
BIMSTEC Tourism Fund	Fund to be set up for the information center to undertake tasks based on the action plan. Each member to contribute $10,000.	The fund was established but has not yet been operationalized. The Colombo meeting recommended the formation of a committee chaired by India to prepare an action plan for using the fund. As of May 2021, six members had sent nominations for the committee.
Tour packages	Tour packages (for two or more countries) including the Buddhist circuit, ecotourism, adventure tourism, and meetings/incentives tourism to be finalized by private stakeholders and promoted among member countries. The tourism focal point of each member country to facilitate preparation of tour packages.	BIMSTEC tour operators organized and offered a variety of multi-country itineraries, many of which were marketed by overseas operators. From March 2020 to May 2021, cross-border travel was still very restricted due to COVID-19. Many tour operators have had to close their businesses. Extensive recovery work will be needed.
Organizing FAM trips	Familiarization trips (two for each member country) for journalists and tour operators to be organized by each member country.	Familiarization trips were standard offers pre-COVID, but post pandemic familiarization trips will initially be on a single country basis due to COVID-19 protocols and restrictions differing from country to country.
Travel facilitation	Expert meeting on a BIMSTEC Business Travel Card to simplify travel visas and immigration procedures.	This did not happen, but post pandemic variations of this idea, especially in the form of vaccine passports are expected and will require regional coordination.

continued on next page

Table 24 continued

Action Title	Proposed Action	Status
Student exchanges	Member countries to facilitate student exchanges.	This did not happen at a BIMSTEC level. Post pandemic, this will depend on COVID-19 restrictions among countries.
Parity in entrance fees at archaeological sites	Parity for nationals of BIMSTEC countries on entrance fees for archaeological sites.	India has initiated this for some archaeological sites.
Extend accessibility by air, land, and water.	Member countries to facilitate more access.	This was happening on a bilateral basis between countries stimulated by the growth in low-cost carriers and cruise lines. COVID-19 put all of this on hold with most flights and cruises drastically reduced and returning very slowly as of May 2021.
Joint investment promotion	Nepal to compile information on investment opportunities and incentives. Member countries to provide information to the Nepal focal point.	Not yet operationalized. During COVID-19 and post pandemic, countries are expected to only focus on investment and business in their country.
Human resource development	Member countries to share information on tourism training facilities via the information center.	Not yet operationalized.
Crisis management	Nepal to work out the operational modalities of the regional network on crisis management.	This has not been operationalized, but because most BIMSTEC countries continue to deal with COVID-19, this is a high priority.
Support from development partners	ADB to examine the proposal of technical assistance, and the Tourism Information Center will follow-up with ADB.	ADB has followed up with BIMSTEC since the center has not yet been set up.
BIMSTEC Tourism Working Group	Establish the group to decide on implementation of the action plan, including use of the fund.	The first working group meeting of the was held on 23 September 2013 in Delhi. The second meeting was postponed many times. It was proposed to be held on 30 May 2021 but was postponed due to a surge in COVID-19 cases in member countries. The group is expected to become more active. Terms of reference have not yet been drafted.
Bangladesh to host the Third BIMSTEC Tourism Ministers Roundtable	Roundtable and workshop in November 2007.	This did not occur in November 2007 in Bangladesh. Postponed many times. Proposed for 31 May 2021 but postponed due to surge in COVID-19 cases.

ADB = Asian Development Bank, BIMSTEC = Bay of Bengal Initiative for Multi-Sectoral Technical and Economic Cooperation.
Source: BIMSTEC Secretariat and Asian Development Bank.

Sample Circuit Itineraries

- ramayana circuit for India, Nepal, and Sri Lanka;[63]
- buddhist circuit for Nepal and India;
- himalaya circuit between Nepal and Bhutan;
- river cruise circuit between Bangladesh and India;
- ocean cruise circuit between Sri Lanka and Kerala, India;
- heritage site circuit from Thailand to Myanmar, starting from Viet Nam; and
- India's Swadesh Darshan circuit program.

Ramayana Circuit for Sri Lanka, India, and Nepal

The Ramayana circuit, which includes destinations in India and Nepal, attracts visitors who want to follow in the path of Lord Rama and Goddess Sita, major deities in Hinduism. In Nepal, the chief destination on the circuit is Janakpur, which was ruled by King Janak, and it is believed that Goddess Sira was born and spent her life there before marrying Rama. Both destinations are just north of the border with Bihar, India.

In India, the circuit includes destinations in nine states and is part of the Swadesh Darshan Scheme of the Ministry of Tourism. Infrastructure improvements are needed in some destinations, including Ayodhya, Nandigram, Shringverpur and Chitrakoot (Uttar Pradesh); Sitamarhi, Buxar, and Darbhanga (Bihar); Chitrakoot (Madhya Pradesh); Mahendragiri (Odisha); Jagdalpur (Chattisgarh); Nashik and Nagpur (Maharashtra); Bhadrachalam (Telangana); Hampi (Karnataka); and Rameshwaram (Tamil Nadu). The Indian Railway Catering and Tourism Corporation had started to offer Ramayana circuit tours just before the outbreak of the COVID-19 pandemic. The Parliamentary Standing Committee on Transport, Tourism and Culture reported on 27 July 2021 that only one of the 15 thematic circuits under the Scheme had been completed and the physical progress of the remaining 14 circuits ranged from 16% to 90%, "which is far from being satisfactory."[64]

Sri Lanka's Ramayana circuit tours of 8 days and 7 nights include the Munneswaram and Manawari Temples in Chilaw, the Thirukoneswaram Temple in Trincomalee, the ancient city of Sigiriya, the Sita Amman Temple in Nuwara Eliya and nearby Ashoka Vatika, the Sri Bhaktha Hanuman Temple in Ramboda, and several other significant religious destinations throughout the country.

Buddhist Circuit for Nepal and India

The Buddhist circuit includes destinations in every BIMSTEC country, but destinations in northern India and southern Nepal are some of the most popular for visitors, especially Buddhists. In India, the main destinations include Shravasthi, where Buddha stayed and expanded his teachings; Kushinagar, where Buddha preached his last sermon and passed away; Vaishali, which has multiple sites of significance to Buddhism; Nalanda, a center of Buddhist studies; Ragir, where Buddha delivered many sermons; Bodh Gaya, where Siddhartha found enlightenment and is considered the spiritual home of Buddhism; and Sarnath, where Buddha delivered his first sermon.

---

[63] A direct bus service was established in 2018 between Janakpur in Nepal, the birthplace of the Hindu Goddess Sita, to Ayodhya in Uttar Pradesh in India.

[64] Parliament of India. 2021. *Potential of Tourist Spots in the Country: Connectivity and Outreach.* Report No. 295. Delhi. Presented to Parliament 27 July 2021.

Buddhist circuit destinations in Nepal include Kapilvastu, where Siddhartha spent his first 29 years and nearby Lumbini where he was believed to have been born, both just north of the Indian border; Devadaha, the birthplace of the Buddha's mother; and Ramgram, home to the Ramagrama stupa, a Buddhist pilgrimage site.

Himalaya Circuit between Nepal, India, and Bhutan

The Himalaya circuit of Nepal, Sikkim in West Bengal, and Bhutan offers extensive cultural, ecotourism, and adventure travel experiences across the country. Itineraries usually start in either Nepal or Bhutan. Starting in Bhutan, first visits can include the fortress of Ta Dzong, which houses the national museum, a hike to the Tiger's Nest Monastery, and the Punakha Dzong, and continue east or west to visit other cultural and nature-based sites. More adventurous visitors can continue further into western and southwestern Bhutan, across Sikkim and into eastern Nepal for treks into the countryside and mountains or vice versa. Most itineraries that combine Nepal and Bhutan involve flights between Kathmandu and Thimpu (Bhutan) to save time. The Sikkim and West Bengal regions between the two countries are rich in travel and tourism experiences for visitors approaching from either Nepal or Bhutan.

In eastern Nepal, some of the destinations offered by the Great Himalayan trail include the Kanchenjunga Base Camp at Pangpema, Makalu Base Camp, and Chhukung in the Everest region. Trails extend north westward covering all of Nepal. The full list of Great Himalaya Treks is listed in Figure 11.

River Cruise Circuit between Bangladesh and India

The river cruise circuit between Bangladesh and India was launched in April 2019, when the MV Mahabaahu, a small cruise ship, left Pandu port in Assam carrying 30 passengers for a 17-day voyage. The itinerary included three UNESCO World Heritage Sites, the Manas National Park in Assam, the Mosque City of Bagerhat in Bangladesh, and the Sundarbans mangrove forest in Bangladesh. A similar cruise was scheduled by Assam Bengal Navigation from Kolkata to Dhaka on the ABN Charaidew II, which was due to visit the Sundarbans mangrove area, the Bhagabatpur crocodile center, Mongla port, Bagerhat mosque, Kochikhali forest station, Katka, Armanik, the guava floating market at Bhimruli, Barisal, Narayangari, and Dhaka.

Figure 11: Full List of Great Himalaya Treks

GHT Treks

GHT Nepal High Route

GHT East Nepal Trek

GHT Central Nepal Trek

GHT West Nepal Trek

GHT Nepal Short Treks

GHT East Bhutan Trek

GHT Central Bhutan Trek

GHT West Bhutan Trek

Annapurna circuit, Naar, Phu

Annapurna Sactuary

Darchula to Rara

Everest Base Camp and Passes

Gosainkund Lake

Himachal Pradesh

Kanchenjunga Base Camp

Karnali Corridor Trek

Kashmir and Ladakh

Khaptad National Park

Kumaon and Garwhal

Langtang Valley

Limi Valley circuit

Makalu Base Camp

Manaslu circuit

Poon Hill and Kopra Ridge

Rara Lake

Rolwaling and Tashi Labsta

Ruby Valley

Tamang Heritage trail

Upper Dolpo circuit

Upper Mustang circuit

GHT = Great Himalaya Treks
Source: Great Himalaya Treks, https://www.greathimalayatrail.com/go-trekking/.

Ocean Cruise Circuit between Sri Lanka and India

Ocean cruises between Sri Lanka and India were increasing in popularity before the COIVID-19 pandemic, especially among Indian tourists. Some of the most popular routes started in Male, Maldives and continued to Colombo, Goa, and Mumbai. A similar itinerary in the opposite direction started in Mumbai and continued to Goa, Managlore, Cochin, Trivandrum (Sri Lanka), and finished in Male. Once travel normalizes, Sri Lankan ports are expected to be especially popular since the government was trying to promote the country's ports as excursion stops.

Heritage Site Circuit from Thailand to Myanmar

Before COVID-19 and political instability hit in Myanmar, the country was growing quickly as an international tourist destination. In 2019, international arrivals were up 23% from 2018. Itineraries often included both Thailand and Myanmar—and to a lesser extent Cambodia, the Lao PDR, and Viet Nam. In Thailand and Myanmar, a circuit focused on visits to heritage sites was a favorite for visitors. Starting in Yangon, one popular itinerary includes visits to Karaweik Royal Barge and the Kyaukhtatgyi pagoda in the city, continuing to Inle Lake to visit the floating village of Ywarna, the Phaung Daw Oo pagoda, Inpawkhan weaving village, and a hot air balloon trip over the pagodas of Bagan. After Myanmar, the tour continued to heritage sites in Bangkok, Chiang Rai, and Chiang Mai.

India's Swadesh Darshan Thematic Circuit Program

India's Swadesh Darshan thematic circuit program could have been the largest both in BIMSTEC countries and Asia and the Pacific if it had proceeded as originally intended. The program offered 76 circuits, which are listed in Annex 3, but, as reported by India's Parliamentary Standing Committee on Transport, Tourism and Culture on 27 July 2021, "only one of the 15 thematic circuits under the Scheme had been completed and the physical progress in respect of the remaining 14 circuits ranges from 16 percent to 90 percent which is far from being satisfactory."[65] Understanding the reasons for the circuit's lack of success warrants a separate report—and perhaps different program design.

Key Features of Other Asian Region Plans and Strategies

The other regional groupings in East Asia and South Asia can offer lessons and best practices for the Plan of Action for Tourism Development and Promotion for the BIMSTEC Region and eventual development of a BIMSTEC regional tourism strategy.

South Asia Subregional Economic Cooperation

The South Asia Subregional Economic Cooperation (SASEC) brings together Bangladesh, Bhutan, India, Maldives, Myanmar, Nepal, and Sri Lanka in a project-based partnership to promote regional prosperity, improve economic opportunities, and a better quality of life in the subregion. SASEC countries share the common vision of boosting intraregional trade and cooperation in South Asia and developing connectivity and trade with Southeast Asia.

The SASEC Tourism Plan, aims to leverage natural resources, promote tourism industry linkages for regional value chains, and expand trade and commerce in South Asia by developing subregional gateways and hubs. ADB, in 2004, sponsored the Human Resource Development and Capacity Building in Tourism Project and the SASEC Tourism Development Project in 2006. The latter focused on leveraging the tourism assets of the SASEC countries.

---

[65]   Footnote 68.

The SASEC Tourism Plan resulted from the second annual meeting of the SASEC Tourism Working Group, which agreed to build on the tourism plans of member countries and establish a thematic framework for future tourism development, and that planning should start with two common themes: ecotourism based on natural and cultural heritage, and Buddhist circuits. The strategic directions of SASEC's tourism plan includes:

- supporting tourism that is sustainable and contributes to reducing poverty,
- keeping to a communication branding that focuses on SASEC's products and not on the SASEC subregion itself,
- initiating joint marketing before introducing measures to ensure product quality,
- promoting the subregion as a tourist-friendly destination,
- facilitating the development of a more competitive tourism industry, and
- improving tourism links with neighboring countries.

To achieve these objectives, the tourism plan presented seven subregional programs and 23 projects, including for the Buddhist circuit. The plan states that other product opportunities involving two or more SASEC members will be promoted.

The World Bank reviewed the plan and subsequent SASEC tourism development activities up to 2012. The plan was updated in 2008 and focused on a simplified framework based on a 5-year action plan with an estimated budget of $452.3 million (Figure 12).[66]

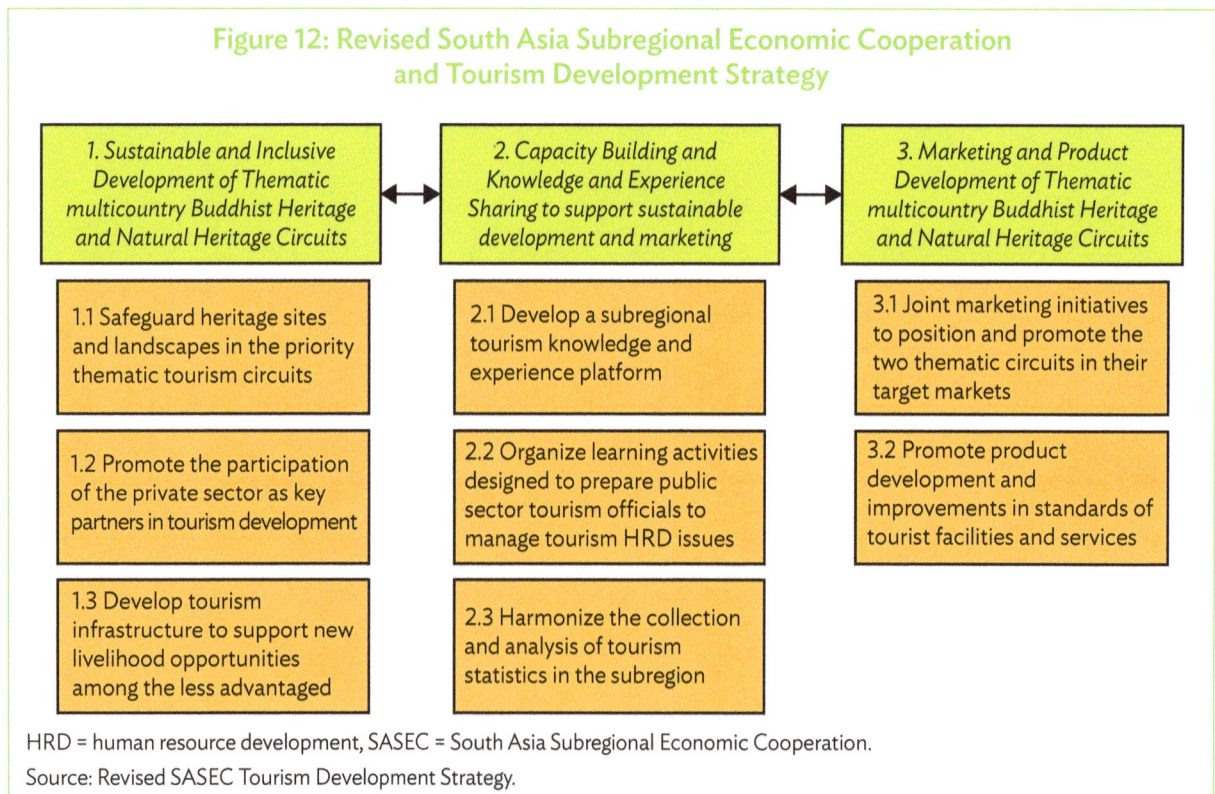

Figure 12: Revised South Asia Subregional Economic Cooperation and Tourism Development Strategy

1. Sustainable and Inclusive Development of Thematic multicountry Buddhist Heritage and Natural Heritage Circuits	2. Capacity Building and Knowledge and Experience Sharing to support sustainable development and marketing	3. Marketing and Product Development of Thematic multicountry Buddhist Heritage and Natural Heritage Circuits
1.1 Safeguard heritage sites and landscapes in the priority thematic tourism circuits	2.1 Develop a subregional tourism knowledge and experience platform	3.1 Joint marketing initiatives to position and promote the two thematic circuits in their target markets
1.2 Promote the participation of the private sector as key partners in tourism development	2.2 Organize learning activities designed to prepare public sector tourism officials to manage tourism HRD issues	3.2 Promote product development and improvements in standards of tourist facilities and services
1.3 Develop tourism infrastructure to support new livelihood opportunities among the less advantaged	2.3 Harmonize the collection and analysis of tourism statistics in the subregion	

HRD = human resource development, SASEC = South Asia Subregional Economic Cooperation.
Source: Revised SASEC Tourism Development Strategy.

---

66   World Bank. 2012. Case Study 1: ADB South Asia Subregional Economic Cooperation Tourism Development Project. http://documents1 .worldbank.org/curated/en/128481563447559559/pdf/ADB-South-Asia-Subregional-Economic-Cooperation-Tourism-Development -Project-Case-Study-1.pdf.

The strategic directions are similar to those that BIMSTEC countries agreed on in their 2005 and 2006 tourism action plans. Some notable features of the SASEC plan that could be considered for inclusion in the updated BIMSTEC plan include:

- building on existing tourism plans in each member country,
- emphasizing safeguards for heritage sites and landscapes,
- establishing a tourism knowledge and experience platform, and
- harmonizing the collection and analysis of tourism statistics in the subregion.

Greater Mekong Subregion

The GMS comprises Cambodia, the Lao PDR, Myanmar, the PRC's Yunnan Province and Guangxi Zhuang Autonomous Region, Thailand, and Viet Nam—all united by the Mekong River. Sponsored by ADB, the GMS Tourism Sector Strategy was launched in 2016 and extends to 2025 to bring GMS countries together to cooperate on regional tourism cooperation. It provides a framework to guide cooperation between national tourism organizations and industry stakeholders. The strategy shares objectives with other tourism plans in the region, such as ASEAN's Tourism Strategic Plan 2016–2025.

The strategy's vision is that "tourism in the Greater Mekong Subregion is integrated, prosperous, equitable, and resilient, with effective partnerships and knowledge management."[67] The strategy's outcome is for a "more competitive, balanced and sustainable destination development." The core strategic directions are human resources development, improving tourism infrastructure, enhancing visitor experiences and services, creative marketing and promotion, and facilitating regional travel. Each of these strategic directions is supported by multiple programs and related activities. The GMS Tourism Sector Strategy describes these programs in detail and sets out estimated budgets. The strategy could provide best practice examples for BIMSTEC. The strategy includes a map showing transport corridors between member states. The corridor stretching from east to west from Viet Nam through the Lao PDR, and Thailand to Myanmar might be the basis for a thematic circuit.

The following are features of the GMS Tourism Sector Strategy that could be considered for inclusion in the updated BIMSTEC Tourism Action Plan:

- Human resource development includes activities to support capacity building for public officials to improve their effectiveness as destination managers, promote entrepreneurship and expand economic opportunities for youth, particularly girls. The short-term training courses offered could be helpful, especially since they include topics needed by BIMSTEC members—such as community-based tourism; fostering ecotourism; cultural and natural heritage interpretation; urban planning; investment promotion; and the use of information technology for destination marketing.[68]
- Infrastructure improvements, such as standardized tourism signage, lighting, and landscape features, rest stops and market facilities in suitable locations alongside roads to serve self-driving and group tours.[69]
- Adopting UNESCO Cultural Heritage Specialist Guide Training Program, especially for circuits linking UNESCO World Heritage Sites.
- Priority thematic multicounty tour programs relevant for BIMSTEC include:

---

[67] Mekong Tourism Coordinating Office. 2017. Greater Mekong Subregion Tourism Sector Strategy, 2016–2025. Bangkok. p. 26.
[68] Greater Mekong Subregion Tourism Sector Strategy 2016–2025. Bangkok, p. 31.
[69] Greater Mekong Subregion Tourism Sector Strategy 2016–2025. Bangkok, p. 34.

- The Southern Coastal Corridor: Cambodia, Myanmar, Thailand, and Viet Nam for beach and islands, leisure, seafood, history, culture, community-based tourism
- The Mekong Tea Caravan Trail: The Lao PDR, Myanmar, the PRC's Yunnan Province, and Thailand for ethnic-group experiences, culture, and ecotourism.
- Promoting the adoption of energy efficiency standards by hotels, and resource-efficient building design and certification programs.
- Producing content to promote multi-country tour programs and experiences, using mobile and fixed digital communication channels and print media, as done by GMS national tourism organizations, the Mekong Tourism Coordinating Office, industry associations, and private sector media partners.
- Promoting multi-country sporting events; in the GMS, these include the annual Mekong Triathlon, cross-border marathons and, potentially, a Tour de Mekong cycling race.
- ADB is the GMS program's impartial coordinating institution for the program and serves as the subregional secretariat.

South Asian Association for Regional Cooperation

The South Asian Association for Regional Cooperation (SAARC) comprises Afghanistan, Bangladesh, Bhutan, India, Maldives, Nepal, Pakistan, and Sri Lanka. Its secretariat was set up in Kathmandu in 1987. The organization has had a Working Group on Tourism since 2004 and has tourism as an area of cooperation. The working group last met in November 2015, when it recommended implementation of the action plan adopted in 2006. An updated plan exists but has not been posted on SAARC's website.

The action plan focuses on the following areas:

- Promoting South Asia as a common tourist destination via
  - annual meetings of the Working Group on Tourism.
  - special events organized by airlines in South Asia and diplomatic missions of member countries, and
  - SAARC tourism brochures made available in the seat pockets of state airlines in SAARC countries.
- Increasing private sector involvement via
  - the increased involvement of the Tourism Council of the SAARC Chamber of Commerce and Industry in tourism marketing and promotion, and
  - inviting tour operators to working group and tourism council meetings.
- Developing human resources
  - Information sharing on facilities for human resource development in tourism in South Asia.
- Promoting a South Asian identity through tourism via
  - direct air links among members,
  - the urgent development of road and rail links between member countries,
  - separate counters for nationals of SAARC countries at airports in the region,
  - mutual recognition of national driving licenses,
  - the same entrance fee for nationals of SAARC countries at archaeological sites and other attractions, and
  - simplified visa formalities for nationals of SAARC countries.
- Developing cultural tourism and ecotourism
  - Encouraging cultural tourism, pilgrimage tourism, medical tourism, and ecotourism via shared best practices.

Most of the features of SAARC tourism plan are already included in BIMSTEC plans, aside from the recommendation for having the same entrance fee for SAARC nationals at archaeological sites and other attractions.

Association of Southeast Asian Nations

ASEAN's 10 member states are Brunei Darussalam, Cambodia, Indonesia, the Lao PDR, Malaysia, Myanmar, the Philippines, Singapore, Thailand, and Viet Nam. In 2016, the group agreed on an ASEAN Tourism Strategic Plan 2016–2025 with the following vision: "By 2025, ASEAN will be a quality tourism destination offering a unique, diverse ASEAN experience, and will be committed to responsible, sustainable, inclusive and balanced tourism development, to contribute significantly to the socioeconomic well-being of ASEAN people."[70]

ASEAN members agreed on the following strategic objectives to realize this vision over the plan period:

- Increase the GDP contribution of ASEAN tourism from 12% to 15%.
- Increase tourism's share of total employment from 3.7% to 7%.
- Increase per capita spending by international tourists from $877 to $1,500.
- Increase the average length of stay of international tourist arrivals from 6.3 nights to 8.0 nights.
- Increase the number of accommodation units from 0.51 units per 100 head of population in ASEAN to 0.60 units.
- Increase the number of awardees for the ASEAN tourism standards from 86 to 300.
- Increase the number of community-based tourism value chain projects from 43 to over 300.

To achieve these objectives, the strategy projected annual growth in international arrivals for Southeast Asia of 5.8% from 2010 to 2020, and 4.3% from 2020 to 2030 (Table 25). The plan document does not include a budget for the tourism strategy.

The following features of the ASEAN Tourism Strategic Plan that could be considered for inclusion in the updated BIMSTEC Tourism Action Plan:

Table 25: International Arrivals to Southeast Asia
(% growth)

Region	Projected Growth in Arrivals		Actual International Arrivals (million)		Projections of International Arrivals	
	2010–2020	2020–2030	2013	2020	2025	2030
World	3.8%	2.9%	1,087	1,360	1,569	1,809
Asia and the Pacific	5.7%	4.2%	248	355	436	535
Southeast Asia	5.8%	4.3%	102	123	152	187

Source: Association of Southeast Asian Nations. Tourism Strategic Plan 2016–2025.

- Recommending emphasis on marketing the region as a single destination.
- Implementing regional tourism standards certification system.
- Implementing the "ASEAN Mutual Recognition Arrangement on Tourism Professionals."
- Establishing tourism-investment corridors.
- Creating and implementing a regional human resources development plan for the tourism industry.
- Producing a tourism investment guide along with the Invest in ASEAN website.[71] A more informative version of this website could be developed for the BIMSTEC region.

---

[70] Association of Southeast Asian Nations (ASEAN). ASEAN Tourism Strategic Plan 2016–2025. Jakarta. p. v.
[71] http://investasean.asean.org.

In November 2020, ASEAN member governments held an online summit that resulted in the adoption of an ASEAN Declaration on an ASEAN Travel Corridor Arrangement Framework. The framework provides a possible model for BIMSTEC Travel Corridors. The declaration focuses on business travel among ASEAN member states and underscores the need to establish a common set of predeparture and post-arrival health and safety measures to protect the well-being and safety of citizens. The declaration makes clear that travelers must abide by the public health regulations required by the authorities of receiving countries.

The ASEAN Coordinating Council, supported by the ASEAN Coordinating Council Working Group on Public Health Emergencies, is tasked with coordinating and overseeing the process of the development and operationalization of an ASEAN travel corridor arrangement framework.

BIMSTEC Member Summary Tourism Profiles

A summary of the strengths, weaknesses, opportunities, and threats analysis for each BIMSTEC country is included in Annex 10.

Bangladesh

Tourism Policy and Strategy

In 2010, the government of Bangladesh introduced a new National Tourism Policy, which sets a tourism vision and mission to be implemented by the Ministry of Civil Aviation and Tourism and its agencies, the Bangladesh Parjatan Corporation and Bangladesh Tourism Board. The board was specifically established under the 2010 Bangladesh Tourism Board Act for tourism sector development, promotion, and marketing. Commenting on the inception report, the Ministry of Aviation and Tourism reported that "following the National Tourism Policy 2010 and other relative rules and regulations and Tourism Vision 2020 guidelines, the government [had] already taken some integrated tourism development projects to promote tourism in Bangladesh ... all these projects are not completed yet, but some are in progress." These projects include:[72]

- a comprehensive tourism master plan, which was expected to be completed by 2021 and drawn up by the Bangladesh Tourism Board;
- establishing tourism facilities in the Sundarbans area and other locations;
- setting up an Exclusive Tourist Zone in Sabrang, Cox's Bazar;
- setting up the Sheikh Hasina Tower and Tourism Zone in Khurushkul, Cox's Bazar;
- building international standard hotels and tourism facilities at Khulna and Sylhet; and
- building tourism facilities near the Paira deep seaport at Patuakhali.

The Bangladesh Tourism Development Plan Tourism Vision 2020 was prepared in 2002. In 2019, a tender was awarded for an updated plan to IOS Partners company.

---

[72] Letter from the Ministry of Foreign Affairs conveying comments from the Ministry of Civil Aviation and Tourism on the ADB Inception Report on Regional Cooperation for Tourism Development in the BIMSTEC Region. 30 July 2020.

The main government guidance for tourism development in Bangladesh has been via the strategy of the Ministry of Civil Aviation and Tourism, which is part of a larger document by the Ministry of Finance. The Ministry of Civil Aviation and Tourism Strategy is Chapter 41 of a larger document that appears to be published by the Ministry of Finance. A request was pending for the full document. Chapter 41 provides details on actions in both civil aviation and tourism. It specifies the major functions and three strategic objectives of the ministry, one of which focuses on the "expansion of tourism." The chapter also highlights the roles and responsibilities of the Ministry for Women's Advancement and Rights. The chapter lists recommendations for future activities.

Another strategy document, Tourism Vision 2020, is mentioned in the ministry's feedback on the inception report and was reported in a trade newspaper article. A copy of this report has been requested.

Thematic Circuit–Based Tourism in Bangladesh

In 27–28 October 2015, the government organized a conference with UNWTO, entitled Developing Sustainable and Inclusive Buddhist Heritage and Pilgrimage Circuit in South Asia's Buddhist Heartland. At the conference, Rashed Khan Menon, who was then the minister for civil aviation and tourism, said this was the first time Bangladesh had been included in the tourism circuit of Buddhist sites of Bhutan, India, Nepal, and Sri Lanka. The conference adopted recommendations for a "comprehensive regional Buddhist circuit road map."[73] Since the conference, however, the roadmap does not appear to have been institutionalized as part of the country's tourism strategy.

The potential for a Buddhist circuit is strong. The country is home to several important Buddhist sites, including Vihara Paharpar, an 8th century temple complex that was once an intellectual center of Buddhism (Figure 13). The complex inspired the Buddhist sites of Borobudur (Indonesia), Bagan (Myanmar), and Angkor Wat (Cambodia). From Dhaka, it is a 9-hour drive—260 kilometers (km) by car or bus—via Bogura, Mahasthangarh to Paharpur (location of the Buddhist Vihara ruins). From Dhaka to Rajshahi, it is possible to travel by bus (three times daily, roundtrip $100) and continue by road from Rajshahi to Paharpur. The drive from Rajshahi airport to Paharpur takes about 4 hours (100 km). By train, several trains connect North Bengal with the Kamalapur Railway Station in Dhaka daily. The closest train station to Parharpur is Jamalganj, Naogaon, which is 5 km away from the site.

Infrastructure improvements at these sites are either planned by the government and in progress. According to the Ministry of Civil Aviation and

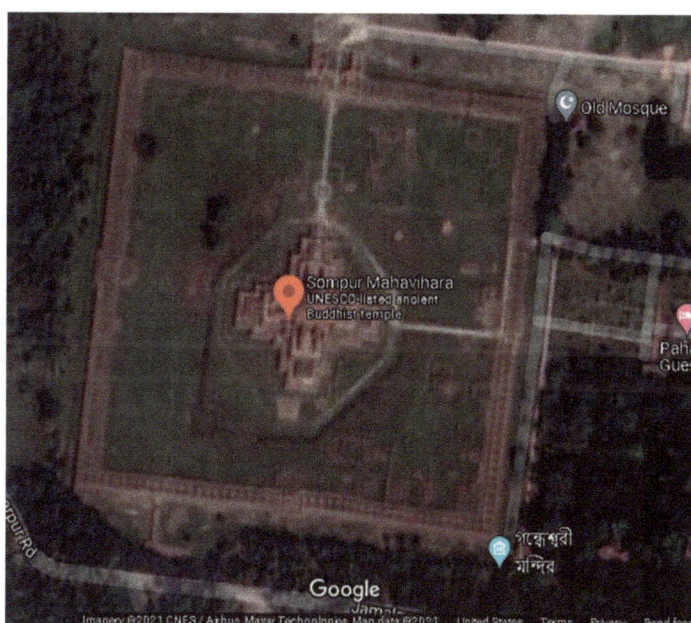

Figure 13: Ruins of the Buddhist Vihara at Paharpur, Bangladesh

Source: Google Maps.

73 UNWTO. 2015. Press release for conference on Developing Sustainable and Inclusive Buddhist Heritage and Pilgrimage Circuit in South Asia's Buddhist Heartland, Dhaka, Bangladesh. https://tinyurl.com/ycttrw7o.

Tourism strategy document, the Bangladesh Tourism Corporation has introduced tourist facilities in the areas of Kantaji temple, Perki, Chottogram, Durgasagar of Barisal district, places adjacent to Sheikh Hasina bridge in Chapainawabganj district, Sirajganj district, Munshiganj of Shatkirah district and Chapti Haor at Dirai Upazila, Sunamganj, Barisal city and Jaflong area. The strategy also lists numerous other infrastructure and product improvements, including an international standard tourism center in Cumilla on the Dhaka–Chittagong highway and a tourism center at Khaliya Juri, Birishiri, and Netrokona. Rest houses with modern facilities have been built in Dhaka, Chittagong, Narail, Rajshahi, and Khulna.[74]

Taslim Amin, executive committee member of Tour Operators Association of Bangladesh, presented several tourism site opportunities in Bangladesh at the October 2015 UNWTO conference. These included archeology, heritage, and historical sites of religious interest (Buddhist, Hindu, and Muslim), such as the Paharpur, Mainamoti, Dhakesshari temples, Ahsan Manzil, and the Star Mosque.[75]

The 708 km river cruise between India and Bangladesh that was proposed in 2019 by the India-based luxury cruise and train tour operator Exotic Heritage Group is another possible circuit. Other potential circuits include linking Bangladesh to the India–Nepal–Bhutan Buddhist circuit.

Coastal tourism is highlighted in the Perspective Plan of Bangladesh, 2021–2041 to establish joint coastal programs with neighboring countries, promote dolphin and whale watching tour packages using tourism as a local corporate incentive program for performance, provide investment and tax incentives to operators, and focus on training professional tour guides. These are all positive signs for tourism, particularly for leveraging circuits as product offers.

Bangladesh has tremendous nature-based and cultural assets that offer opportunities for developing tourism, especially circuit-based tours and segments that overlap with other circuits, such as community-based tourism, adventure travel, and river tourism.

Tour Operators in Bangladesh

Several Bangladesh tour operators offer itineraries that include the Paharpur complex; among them are Global International BD Tours and Travels, JourneyPlus, Mam Holidays, PathFriend, ShareTrip, and Tourist Channel Bangladesh.

The survey of 34 respondent Bangladeshi tour operators conducted in the summer of 2020 found 25 (73.5%) saying they offered circuits to international or domestic tour groups. While most of these offers were for international visitors (67.7%), nearly a quarter of respondents said they offered circuits to domestic groups. Respondents said they expected the domestic market to increase for the rest of 2020 due to COVID-19 restrictions on international travel. As Figure 15 shows, most of respondents (76%) were "very optimistic" before COVID-19 that the domestic market would increase. This bodes well for a near-term market focus on local and domestic markets.

COVID-19 Measures to Help the Tourism Industry Recover

Government assistance will be critical to help the tourism industry recover from the economic impact of the COVID-19 pandemic. The Bangladesh Tourism Board, in its guidelines for reopening the industry, noted that the pandemic was a "catastrophe" for the industry and that "people from all walks of life suffered." In addition,

---

74   Government of Bangladesh, Ministry of Civil Aviation and Tourism. National Tourism Strategy, Chapter 41.
75   Report of the First Meeting of BIMSTEC Network of Tour Operators. 7 July 2017. Delhi.

"about 4 million tourism workers and at least 1.5 million dependents plunged into uncertainty," said the board.[76] The following were among the measures taken by the government to assist the travel and tourism industry:

- Stimulus packages providing capital to help the industry recover.
- Government-subsidized loans through commercial banks at an interest rate of 4.5%.
- Setting up the Committee for Crisis Management of Tourism Industry under the Ministry of Civil Aviation and Tourism, which works on mitigating the effects of COVID-19.[77]
- The Bangladesh Tourism Board's COVID-19 Recovery Plan.[78]
- Standard operating procedures for restarting tourism that have guidelines for each segment; these comply with UNWTO, World Health Organization, and Ministry of Health and Family Planning regulations.

Bhutan

Tourism Strategy and Policy

In 2016, the National Council of Bhutan's Review Report on Tourism Policy and Strategies reiterated the country's "high value, low volume" approach. This is both the guiding policy on tourism and, since 1974, the model for managed tourism. The approach is also in line with the country's famous Gross National Happiness development philosophy and approach. The report describes the important role of tourism in the country's Eleventh Plan, including the roles played by stakeholder institutions and organizations. It notes that despite the lack of an updated policy or strategy for tourism, the industry has helped accelerate economic growth in a sustainable way.[79]

The Government of Bhutan, in its feedback on the inception report, said:

> Tourism in Bhutan has been guided by the time-tested policy of "High value, low volume" and tourism has been prioritized as one among five jewels as per the Economic Development Policy 2016 which is an apex policy for guiding the economic development in the country.
>
> The tourism strategy and development plans 2013–2018 was implemented as part of the 11th Five-year plan. However, tourism has been accorded flagship status during the 12th Five Year Plan given its importance and potential to contribute to socioeconomic development of the country.
>
> Further, the Government having assessed the need of comprehensive national tourism policy has worked towards formulating the comprehensive policy to re-affirm the guiding principle of "High Value, Low Volume" in 2019, which at present has been endorsed by various committees and is at the highest committee for endorsement and approval.[80]

In October 2019, a new national tourism policy was introduced, which provided for the establishment of the Tourism Council of Bhutan. The policy specified improvements in product and marketing, human resources, and quality standards. It builds on the 2016 Review Report on Tourism Policy and Strategies, which reiterated the country's

---

[76] Bangladesh Tourism Board, Ministry of Civil Aviation and Tourism. 2020. *Standard Operating Procedures to Restart the Tourism Industry during COVID-19.*

[77] UNWTO. 2020. *COVID-19: Measures to Support the Travel and Tourism Sector.* 9 April. https://www.unwto.org/.

[78] Bangladesh Tourism Board. 2020. *Recovery Plan of Bangladesh Tourism Board from COVID 19.* 1 June. https://tourismboard.portal.gov.bd/site/page/7f734bbf-4185-47f3-bd9c-dec79868d54e.

[79] E-mail dated 1 June 2020 from the Tourism Council of Bhutan.

[80] Feedback from the Ministry of Foreign Affairs, Government of Bhutan, 6 July 2020.

Guiding Policy on Tourism for the Economic Affairs Committee. The new national tourism policy also reinforced Bhutan's guiding policy of "high value, low volume" (footnote 77). The policy was endorsed by the government in January 2021.

The updated tourism policy outlines the vision for Bhutan to be a "green, sustainable, inclusive and high value tourism destination to promote High value, Low volume tourism."

The guiding principles are:

- Develop and promote forms of tourism that are consistent with the national development philosophy of Gross National Happiness
- Promote "high value, low volume" tourism
- Promote tourism that does not undermine national security or erode the tangible and intangible cultural heritage and environment
- Promote inclusive and equitable growth
- Ensure sustainable tourism development.

The tourism policy sets the agenda and direction for developing tourism; its success will depend on the integrated and holistic engagement and participation of all stakeholders. The policy document is supported and complemented by a comprehensive strategic development plan and incentive packages to boost sustainable growth.[81] It also outlines reform measures under the following thematic and domain areas:

- Sustainable tourism development
- Positioning tourism in development policies and programs
- Strengthen tourism governance and institutional setup
- Improvement of tourism products and investment environment
- Seasonal and geographical spread of tourism
- Standards and quality
- Human resources
- Partnerships and collaboration
- Inclusive and integrated tourism

For the country's tourism strategy, the government has prioritized tourism as a "Flagship Program," making it an important economic sector with flagship status in the 12th Five-Year Plan, 2018–2023. The strategic framework for the program includes:

a. Objective: Increased contribution from "high value, low volume" tourism to the national economy, while ensuring a more equitable spread of tourism and its benefits.
b. Outcomes:
    - Increased contribution of tourism to the national economy and rural livelihoods.
    - Balanced regional development and seasonal spread of tourism.
    - Enhanced visitor experience of Bhutan as an exclusive destination.
c. Approach:
    - Integrated circuit development
    - Eastern circuit

---

81   Tourism Council of Bhutan. Tourism Policy of the Kingdom of Bhutan. Approved policy. October 2019. pp. 6–7.

- Western circuit
- Central circuit
- Southern circuit
- Enabling environment
- Policy and regulatory frameworks
- Institutional strengthening
- Enhanced service delivery
- Promotion and branding
- Human resources and skills
- Standards and safeguards
- Research and development

Thematic Circuit–Based Tourism in Bhutan

Integrated circuits are part of the country's tourism strategy. The government intends to explore themes unique to the circuits discussed earlier, which are expected to include nature- and culture-based visitor experiences, and wellness, spiritual, and other special interests. These are expected to include the Buddhist circuit in partnership with India, Nepal, and Sri Lanka, and the "Two Kingdoms: One Destination" initiative with Thailand.

In the survey conducted for this report with tour operators in Bhutan, which had 17 responses, 82% said they offered circuits to international and domestic groups. Nearly a third offered circuits to domestic groups, although this is not a priority market, and 88.2% offered circuits to international and/or domestic groups (Figure 14).

Figure 14: Bhutan Tour Operators and Circuits Survey (2020)

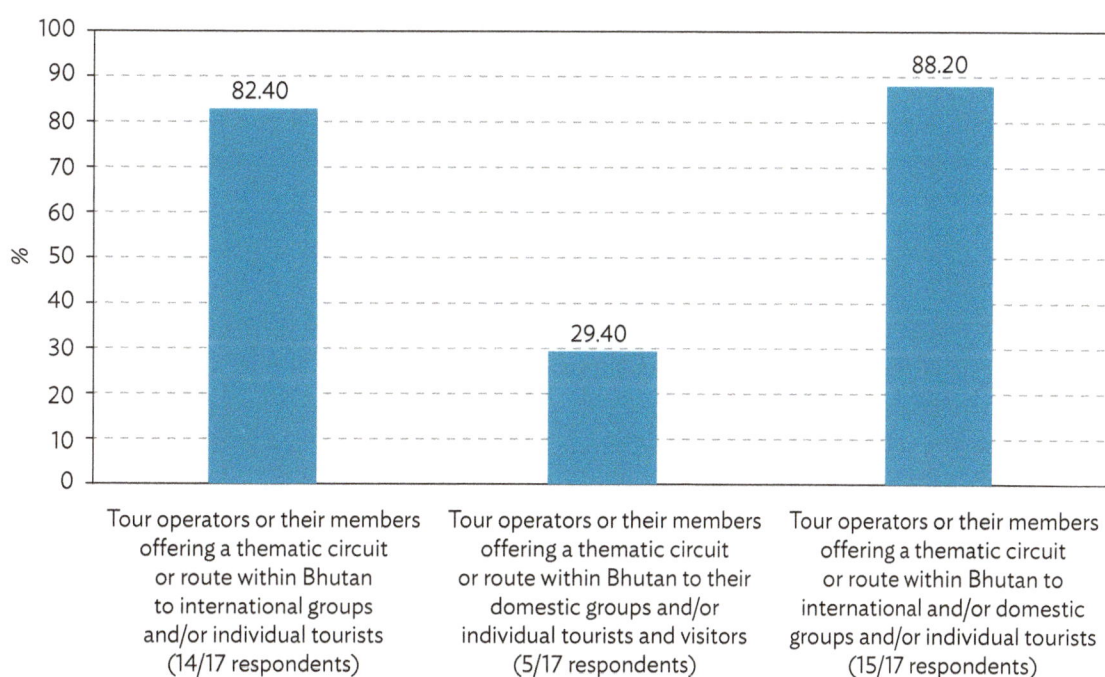

Category	%
Tour operators or their members offering a thematic circuit or route within Bhutan to international groups and/or individual tourists (14/17 respondents)	82.40
Tour operators or their members offering a thematic circuit or route within Bhutan to their domestic groups and/or individual tourists and visitors (5/17 respondents)	29.40
Tour operators or their members offering a thematic circuit or route within Bhutan to international and/or domestic groups and/or individual tourists (15/17 respondents)	88.20

Source: Asian Development Bank.

Tour operators were asked about their level of optimism for increased tourism in Bhutan. Not surprisingly, as the country's brand is "Happiness is a Place," and Bhutan is renowned globally for its Gross National Happiness Index, three quarters of respondents were "very optimistic" and 18% were "optimistic" (Figure 15).

Figure 15: Pre-COVID-19 Optimism Survey of Tour Operators in Bhutan (2020)

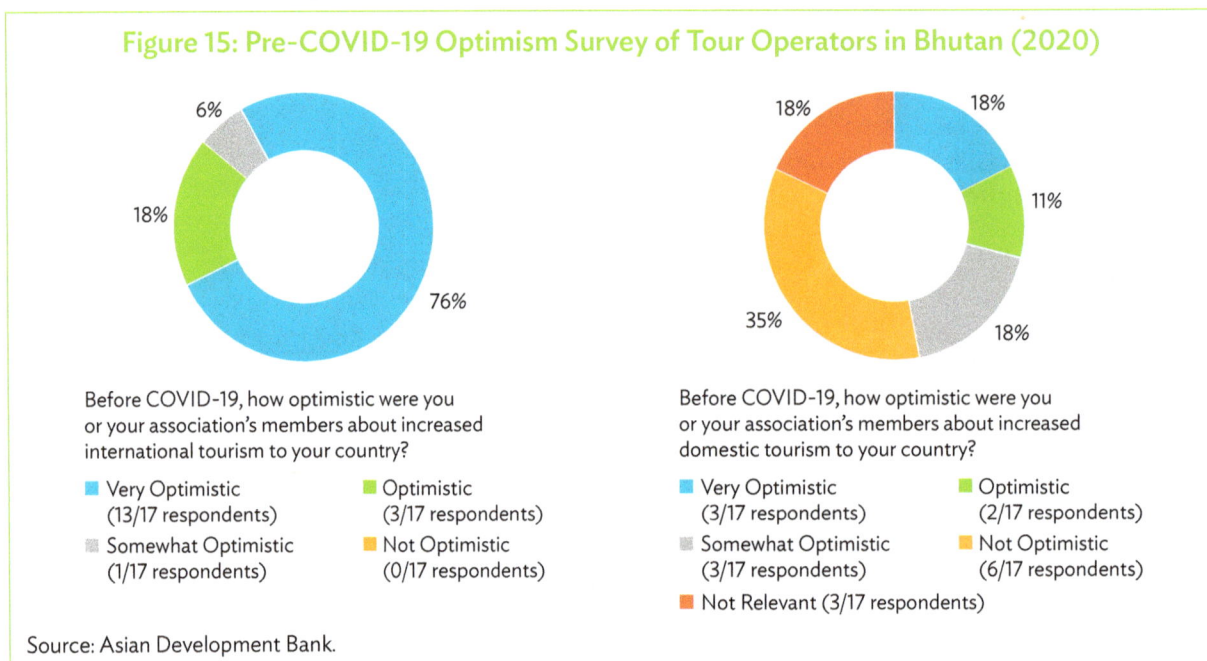

Before COVID-19, how optimistic were you or your association's members about increased international tourism to your country?

- ■ Very Optimistic (13/17 respondents)
- ■ Optimistic (3/17 respondents)
- ■ Somewhat Optimistic (1/17 respondents)
- ■ Not Optimistic (0/17 respondents)

Before COVID-19, how optimistic were you or your association's members about increased domestic tourism to your country?

- ■ Very Optimistic (3/17 respondents)
- ■ Optimistic (2/17 respondents)
- ■ Somewhat Optimistic (3/17 respondents)
- ■ Not Optimistic (6/17 respondents)
- ■ Not Relevant (3/17 respondents)

Source: Asian Development Bank.

COVID-19 Measures to Help the Tourism Industry Recover

The government's Economic Stimulus Plan was formulated to aid industries, including tourism, hit by the COVID-19 pandemic by providing capital to help them recover.[82] Bhutan partially reopened its tourism sector in mid-June 2021, and seemed to be on the path to recovery even though international travel remained restricted. The government has set up a committee to work with stakeholders and a national emergency response team to combat the pandemic and implement recovery plans. Bhutan is working on its 21st Century Economic Roadmap to guide economic development over the next 10 years.

Bhutan and India, in December 2020, agreed on a "travel bubble" arrangement for limited international flights between the countries. For visitors arriving in Bhutan, a negative polymerase chain reaction (PCR) test was required; the test must be done no more than 3 days before arrival. A 21-day quarantine was also required upon arrival.

India

Tourism Strategy

The 2002 National Tourism Policy of India's Department of Tourism, which specifies development goals, objectives, and strategies, has not been updated. Tourism does feature in more recent overall national economic development

---

[82]   UNWTO. 2020. *COVID-19: Measures to Support the Travel and Tourism Sector.* 9 April.

and planning documents. But all national planning and policy efforts are challenged by a lack of coordination with the country's 29 states, each of which has its own tourism policies, plans, and strategies, which are often not synchronized with the work of the central government.

India had 17.9 million international arrivals and $30.7 billion in tourism receipts in 2019, putting it 24th and 13th globally on these measures.[83] Among BIMSTEC countries, India had the highest number of international tourists and receipts after Thailand. Australia and Japan were earning more and receiving more international visitors. For international arrivals, countries with smaller populations than India got more arrivals in 2019, for example, the Netherlands (20.1 million), Portugal (24.6 million), and Malaysia (26.1 million). India's diverse cultural and nature offers boosted international arrivals from 5.8 million in 2010, and international receipts from $14.5 billion in 2010 to $30.7 billion in 2019.[84] Domestic tourism, a vital market for India, increased from 747.7 million visits in 2010 to 2.3 billion in 2019 (Table 15). The states of Andhra Pradesh, Karnataka, Uttar Pradesh, and Tamil Nadu accounted for over half of all domestic visits.[85]

India's relatively strong performance in tourism has been undermined by weak infrastructure, marketing, and governance, as well as weak cooperation with the public and private sector in Bangladesh, Bhutan, and Nepal to create thematic circuits. In 2019 and 2020, however, the government started pushing for the further development of tourist circuits via financing from programs, including for the Swadesh Darshan Scheme.

Thematic Circuit–Based Tourism in India

The government-led Swadesh Darshan Scheme was the most comprehensive and organized thematic circuit system proposed for any BIMSTEC country. Set up in 2014, it provides funding to each state to establish or improve circuits for the 15 themes shown in Figure 16.

The scheme is a logical follow-on to the successful Incredible India campaign, which helped market and promote the country's rich cultural, historical, religious, and natural heritage. The government recognized that this could be a source of overall economic growth and that heritage-based circuits can be an effective way of developing and marketing tourism.

Special interest circuits are regarded globally as an effective and integrated

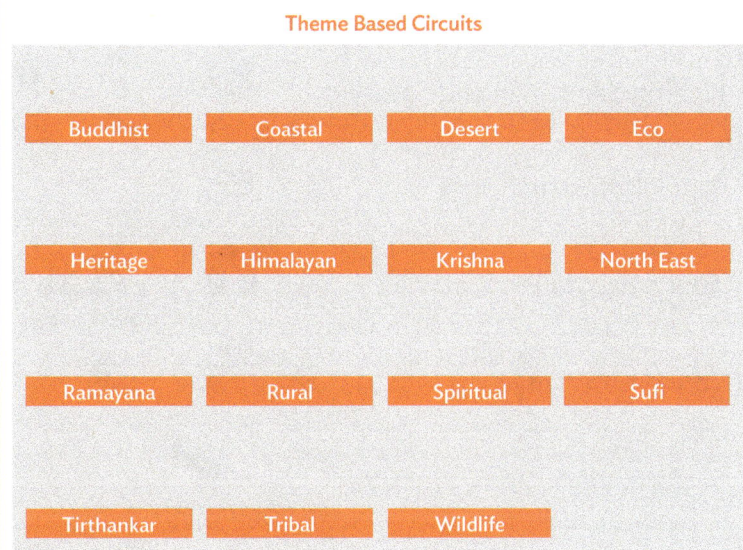

Figure 16: Swadesh Darshan–Themed Circuits Scheme

SDS = [Swadesh Darshan Scheme ?]
Note: Photos for the Sufi and Ramayana circuits were not included with this graphic on the SDS website.
Source: Ministry of Tourism, India, Swadesh Darshan Scheme, www.swadeshdarshan.gov.in.

[83] UNWTO. World Tourism Barometer Statistical Annex. July 2021. Madrid.
[84] UNWTO. World Tourism Barometer Statistical Annex. June 2020. Madrid.
[85] Government of India, Ministry of Tourism. 2019. India Tourism Statistics at a Glance: 2019. Delhi. p. 19.

way to organize especially heritage-based tourist visits. As mentioned above, a tourist circuit is defined by the UNWTO as: "A subsystem of tourism that focuses on the needs of people while geographically mobile, including transport facilities, availability of information, and the proximity of stopover attractions. From the standpoint of the tourist, the act or experience of briefly visiting a number of areas as part of a single round trip."[86]

The Ministry of Tourism says that the Swadesh Darshan Scheme and other theme-based tourist circuits can be "developed in a manner that supports communities, provides employment and fosters social integration without compromising … environmental concerns and provide unique experiences to the tourists."[87]

In mid-August 2020, this report received updated information from the program, which indicated that 739 tourist destinations in 31 states and union territories are covered via 76 Swadesh Darshan projects. The scheme had budgeted ₹5,840 Crore (Cr) ($787 million)[88] for these projects and that ₹3,209 Crore ($432 million) has been disbursed.[89] As Figure 17 shows, the scheme is intended to support multiple types of development, several of which are related to infrastructure (roads, sanitation, and nature trails, for example). Among these areas of support, a third of this amount was dedicated to assisting the creation of core tourism products, nearly a quarter was for trunk infrastructure, and the rest for varying percentages of support.

Figure 17: Swadesh Darshan Scheme Areas of Support

Connectivity	Roads / Pathways Helipad
Health, Sanitation and Safety	First Aid Illumination CCTV Safety Accessibility SWM Sanitation
Trunk Infrastructure	Utility Infrastructure General Site Development Landscaping
Tourism Core Products	Conversation Trails / Camping TFC / Interpretation Center
Tourism Activities	Souvenir Shops Nature Trails Cafeteria
Performing Art Infrastructure	OAT's / Cultural Centers
Soft Intervention	GIS Mapping Mapping Website Development / Promotion Capacity Building

GIS = Geographic Information System, SWM = Swacch Bharat Mission, TFC = Tourist Financial Center.

Note: OAT is undefined in the Toolkit.

Source: Presentation by Deputy Secretary Rajesh Kumar Sahu, Government of India, Ministry of Tourism. 14 August 2020.

---

86   UNWTO Secretariat of State for Tourism of France. 2001. *Thesaurus on Tourism and Leisure Activities*. p. 373.
87   Government of India, Ministry of Tourism. Swadesh Darshan Program. http://swadeshdarshan.gov.in/.
88   ₹74.24 = $1 as of 25 August 2020.
89   Presentation by Deputy Secretary Rajesh Kumar Sahu, Government of India, Ministry of Tourism. 14 August 2020.

The Swadesh Darshan Scheme reported that, as of mid-August 2020, the following five projects had been completed under the scheme, costing ₹361 Cr ($48.6 million):

- Rajasthan (2015–2016): Development of Shakambhari Mata Temple, Sambhar Salt Complex, Devyani Kundu, Sharmistha Sarovar, Naliasar, and other destinations.
- Mizoram (2015–2016): Development of Thenzawl and South Zote-Districts Serchhip– Reiek.
- Manipur (2015–2016): Development of Imphal Khongjom.
- Sikkim (2015–2016): Development of circuit linking Rangpo (entry)–Rorathang–Aritar–Phadamchen–Nathang–Sherathang–Tsongmo–Gangtok–Phodong–Mangan–Lachung–Yumthang–Lachen–Thangu–Gurudongmer–Mangan–Gangtok–Tuminlingee–Singtam (exit).
- Andhra Pradesh (2015–2016): Development of circuit linking Kakinada–Hope Island–Coringa Wildlife Sanctuary–Passarlapudi–Aduru–S Yanam–Kotipally.

A July 2021 report by the Parliamentary Standing Committee on Transport and Tourism found that only one of the Swadesh Darshan Scheme's 15 thematic circuits had been completed and that progress in the other 14 ranged from being 16% to 90% complete—a level, that the committee said, was "far from being satisfactory." Further research on the reasons for the lack of success warrants a separate report and perhaps different program design.

The Swadesh Darshan Scheme's Buddhist circuits is in five states and included, as of mid-August 2020, projects at varying levels of completion. These included the Sanchi-Satna-Rewa-Mandsaur-Dhar itinerary in Madhya Pradesh, the Srawasti-Kushinaga-Kapilvastu itinerary in Uttar Pradesh, building a Cultural Centre at Bodhgaya in Bihar, the Junagadh–Gir Somnath–Bharuch–Kutch–Bhavnagar–Rajkot itinerary and the Shalihundam-Thotlakonda-Bojjanakonda-Amravati-Anupu itinerary in Andhra Pradesh. One of the few success stories appears to be Madhya Pradesh's itinerary, which was reportedly 88% complete.

Some completed Swadesh Darshan Scheme projects have reportedly achieved additional albeit limited success, including the Sambhar Lake Town in Rajasthan on the Desert circuit, which includes the privately operated Sambhar Salt Train (culinary experience, caravan park, and mini-desert safari) and the privately operated Lachen & Tumin Lingee holiday destination (log huts and wayside amenities) in Sikkim on the Northeast circuit.

In the survey, 71 responses were received from Indian tour operators (Figure 18). Fifty-four of them (76.1%) offered circuits to international and/or domestic groups. More than half offered circuits to domestic groups and 69.0% (49) offered circuits to international groups.

Stimulating the development of circuits became a higher priority for the Ministry of Tourism in June 2020 when it reportedly requested state tourism boards to identify itineraries that connect destinations in each state to promote domestic travel. The objective was to encourage more domestic tourism, especially close to home and within short driving distances from cities.

As with the Swadesh Darshan Scheme, the ministry was looking to the states to create itineraries, especially ones that focus on experiential travel in less-visited destinations. The ministry will then circulate the itineraries to the travel sector in India.[90]

90   P. K. Kumar. 2020. Ministry of Tourism Asks States to Chart out New Circuits to Promote Domestic. *ET Travel World*. 22 June. https:// travel.economictimes.indiatimes.com/news/ministry/ministry-of-tourism-asks-states-to-chart-out-new-circuits-to-promote-domestic-travel/76508496.

Figure 18: India Tour Operators and Circuits Survey (2020)

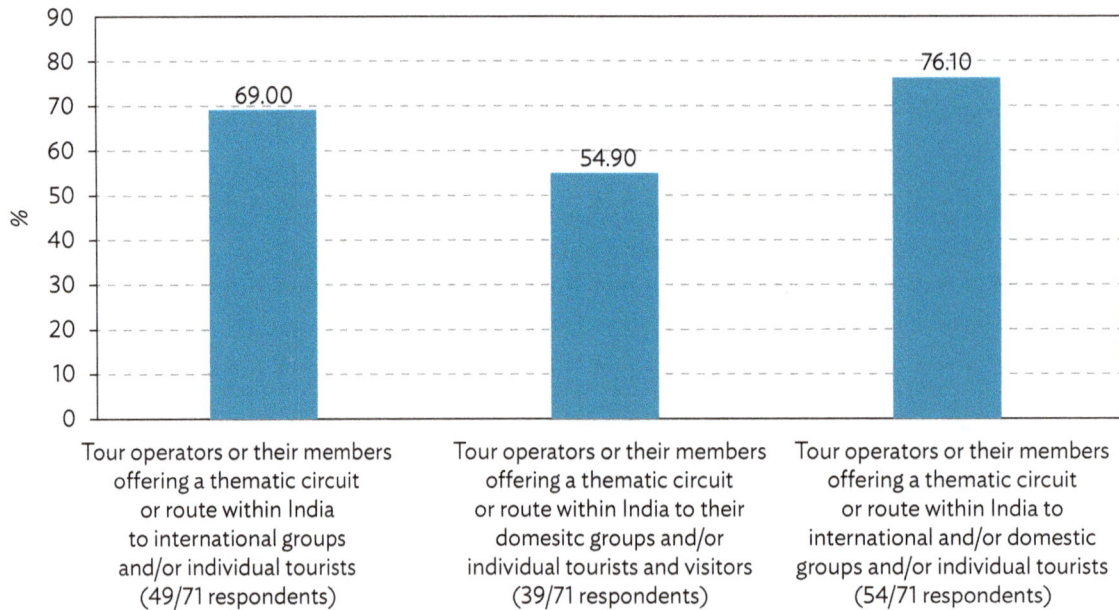

Source: Asian Development Bank.

The domestic tourism initiative saw some success until the outbreak of COVID-19, and this was reflected in survey responses. Asked about their level of optimism for increased tourism to and within India, nearly two-thirds were "very optimistic" (50 out of 71 respondents). Some 30% (21) were "optimistic" about increased international tourism to India. For domestic tourism, 51% (36) were "very optimistic" and 38% (27) "optimistic" (Figure 19).

Figure 19: Pre-COVID-19 Optimism Survey of Tour Operators in India

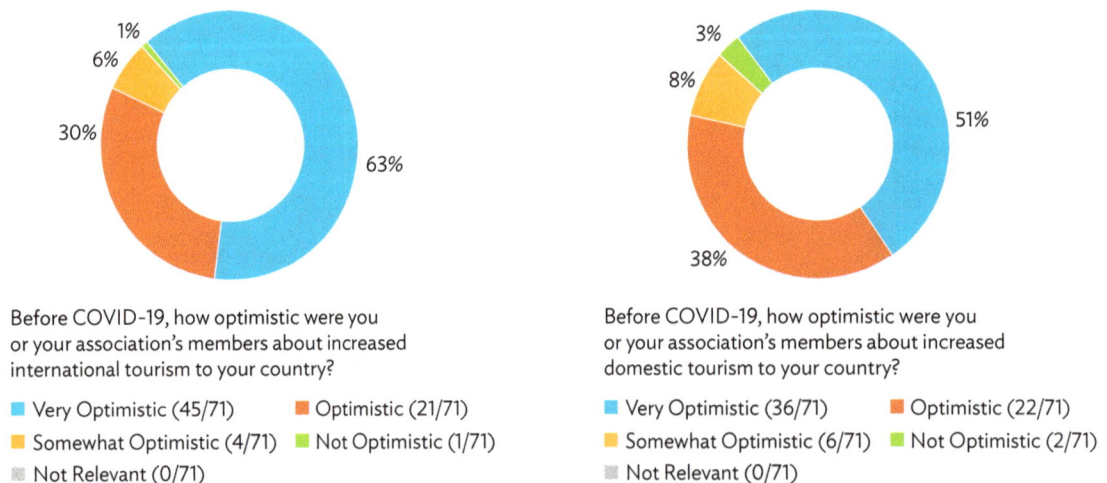

Before COVID-19, how optimistic were you or your association's members about increased international tourism to your country?

- Very Optimistic (45/71)
- Optimistic (21/71)
- Somewhat Optimistic (4/71)
- Not Optimistic (1/71)
- Not Relevant (0/71)

Before COVID-19, how optimistic were you or your association's members about increased domestic tourism to your country?

- Very Optimistic (36/71)
- Optimistic (22/71)
- Somewhat Optimistic (6/71)
- Not Optimistic (2/71)
- Not Relevant (0/71)

Source: Asian Development Bank.

COVID-19 Measures to Help the Tourism Industry Recover

The government focused on domestic and intra-state tourism in 2020 to help restart the industry. With many flights grounded in the country, the government promoted visits to attractions closer to where people lived. The Dekho Apna Desh campaign was part of this effort; it presents videos and photos online of destinations across the country (Figure 20).

Health measures and protocols to control the spread of COVID-19 have been circulated to hotels, other accommodation providers, tour and travel operators, restaurants, and homestay units.

The Incredible India Facilitator Program was launched by the Ministry of Tourism to create a pool of trained professionals for facilitating visits when the country reopens to international tourism.

Figure 20: Dekho Apna Desh Campaign

Source: Ministry of Tourism, India.

Myanmar

Tourism Strategy

The Myanmar Tourism Master Plan, 2013–2020 is the Ministry of Hotels and Tourism's main strategy for development of tourism. The plan calls itself a "roadmap to shape the future of tourism in Myanmar." It is a well-structured document that sets out a clear vision. It has nine guiding principles and six main strategic programs, each with a set of objectives that form a long-term implementation framework. The plan also recognizes the need for "strong coordination" among public and private sector stakeholders to ensure its successful implementation.

Myanmar began opening up to international tourism as political reforms began to take hold from 2010. In that year the country received 791,500 international tourists; by 2019, this had grown to 4.36 million. Over this period, international tourism receipts soared from $254 million to $2.5 billion (Table 6). Visitors are attracted by the country's diverse cultural assets, which include World Heritage Sites, over 100 ethnic groups, 36 protected areas, nearly 3,000 km of coastline, and a rich history manifested in well-preserved religious and vernacular architecture. To help Myanmar tourism, ADB and other donors have been assisting with strategies and plans.

The following Myanmar Tourism Master Plan's objectives correspond to the main challenges facing the country in developing its tourism industry:

- Strengthen the institutional environment
- Build human resource capacity and promote service quality
- Strengthen safeguards and procedures for destination planning and management,
- Create quality products and services
- Improve connectivity and tourism-related infrastructure
- Build the image, position, and brand of Tourism Myanmar.

Table 26: Implementation Status of the Myanmar Tourism Master Plan

Program No.	Strategic Program	No. of Projects	No. of Projects Accomplished	No. of Ongoing projects	No. Remaining Projects	Not Needed
1	Strengthen the institutional environment	4	2	2		
2	Build human resource capacity and promote service quality	10	4	4	1	1
3	Strengthen safeguards and procedures for destination planning and management	8	5	2	1	
4	Develop quality products and services	5	3	1	1	
5	Improve connectivity and tourism-related infrastructure	8	3	1	4	
6	Build the image, position, and brand of Tourism Myanmar	3	1	2		
Total		**38**	**18**	**12**	**7**	**1**

Source: Myanmar Tourism Master Plan.

The ministry was planning to create the next tourism master plan for 2020–2030 based on the 2013–2020 plan in coordination with state and regional governments and the assistance of the Luxembourg Agency for Development Cooperation. Deeper research and inquiry, however, are needed because progress on both was unclear.

Thematic Circuit–Based Tourism in Myanmar

Thematic circuits are not mentioned in the Tourism Master Plan, 2013–2020 or the Myanmar Ecotourism Policy and Management Strategy, 2015–2025. In the survey of tour operators, two-thirds of the 15 respondents from Myanmar said they offered circuits to international and/or domestic groups. Of these, 53.3% (8) offered circuits to international groups and 46.7% (7) offered circuits to domestic groups (Figure 21).

Figure 21: Myanmar Tour Operators and Circuits (2020)

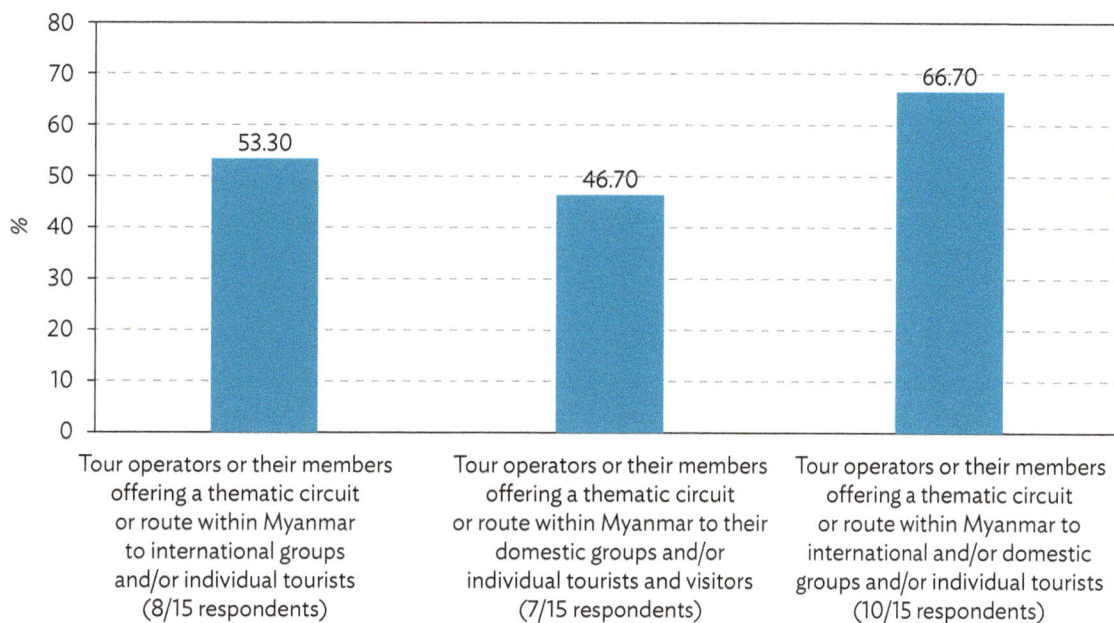

Source: Asian Development Bank.

In the survey of tour operators, 94% of tour operators were "very optimistic" or "optimistic" about the prospects for increased international tourism. For domestic tourism, 30% were "very optimistic" and nearly 50% were "optimistic" (Figure 22).

Myo Myo Than, deputy director, Tourism Promotion Department, Ministry of Hotels and Tourism, at an October 2015 conference with the UNWTO, presented several opportunities for developing tourism. These included the sites of Myeik Archipelago, Natmatung (Mt. Victoria), Hkakabo Razi Kachin, Hpaan Kayin, Mrauk Oo Rakhine, Loikaw Kayah State, and UNESCO World Heritage Sites. Than explained Myanmar's efforts to develop and promote community-based tourism, which is directly relevant for developing thematic circuits. Myanmar is a lead country for developing Buddhist pilgrimage tourism between GMS countries and India.[91]

---

[91]   Report of the First Meeting of BIMSTEC Network of Tour Operators. 7 July 2017. Delhi.

Figure 22: Pre-COVID-19 Optimism of Tour Operators in Myanmar

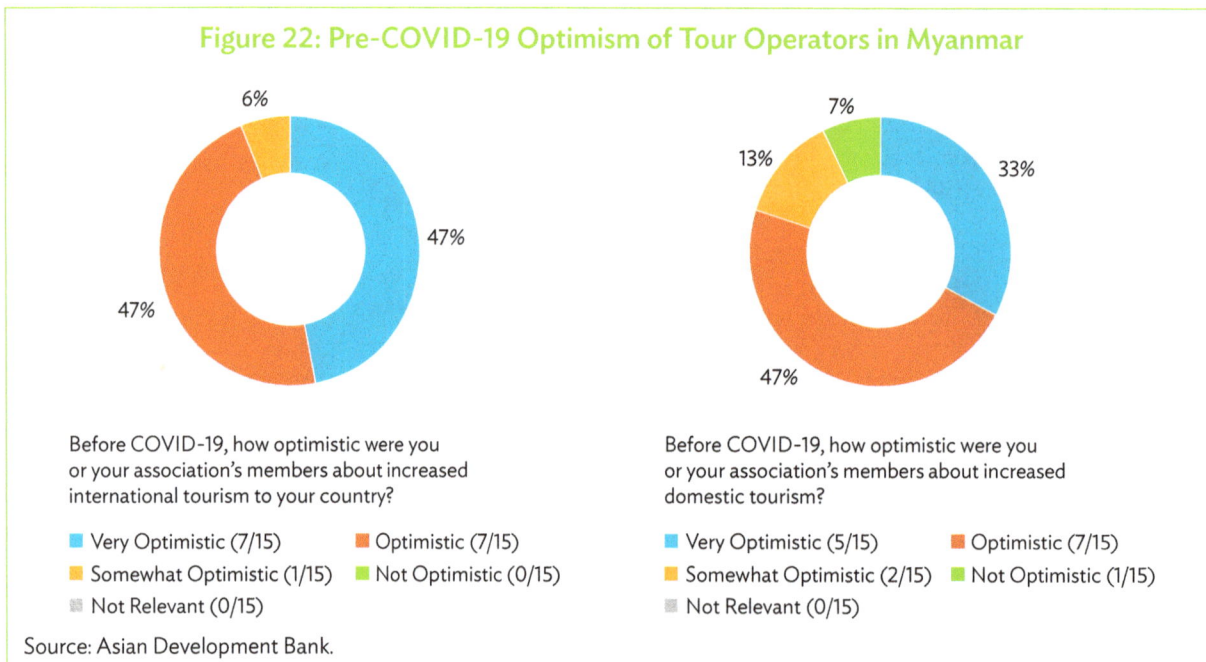

6%

47%

47%

Before COVID-19, how optimistic were you
or your association's members about increased
international tourism to your country?

■ Very Optimistic (7/15)   ■ Optimistic (7/15)
■ Somewhat Optimistic (1/15)   ■ Not Optimistic (0/15)
▨ Not Relevant (0/15)

7%

13%

33%

47%

Before COVID-19, how optimistic were you
or your association's members about increased
domestic tourism?

■ Very Optimistic (5/15)   ■ Optimistic (7/15)
■ Somewhat Optimistic (2/15)   ■ Not Optimistic (1/15)
▨ Not Relevant (0/15)

Source: Asian Development Bank.

COVID-19 Measures to Help the Tourism Industry Recover

The government enacted the detailed COVID-19 Tourism Relief Plan comprised of three strategies in August 2020 and onward, each with four action programs and multiple activities for each program (Table 27).

Table 27: Myanmar COVID-19 Tourism Relief Strategies

Strategies	Programs
Survival (self-finance and stimulus package)	1.1 Reduced taxes, waived license fees and lease fees for hotel and tourism business 1.2 Stimulus package 1.3 Cushioning the impact on tourism professionals and staff 1.4 Evaluating the market and the positions of products and destinations
Reopening (relaxing of lockdown and quarantine)	2.1 Health and safety of travelers and staff 2.2 Conducting paid training programs 2.3 Marketing for "new normal" situation 2.4 Promoting e-commerce platform and digital payment
Relaunching (reinventing Myanmar Tourism and relaxing of regulations), August 2020 to January 2021	3.1 Communication campaign and marketing 3.2 Travel facilitation 3.3 Incentive programs for investment 3.4 Finding grants and loans from development partners.

Source: Government of Myanmar, Ministry of Hotels and Tourism. COVID-19 Tourism Relief Plan. https://tourism.gov.mm/wp-content/uploads/2020/06/COVID-19-Tourism-Relief-Plan.pdf (accessed 15 July 2022).

Plan activities include:

Fiscal measures:
- Stimulus packages
- 6-month tax payment delay
- 2% withholding tax free for export services
- Tax exemptions and loans for hotels and other tourism services
- Special 1% interest credit lines to help businesses continue operating
- Reduced interest rates for commercial loans.

Additional measures:
- Government task force to launch a post-COVID-19 tourism strategy
- The government conducting research and market analysis to help the tourism industry formulate contingency plans and track tourism demand
- Real-time customer service online via "chatbots" to quickly respond to tourist queries
- Online tourism training provided by the Ministry of Hotels and Tourism, and sponsored by tourism employers
- Furloughed tourism professionals hired as trainers.[92]

Nepal

Tourism Strategy

Nepal's Ministry of Culture, Tourism and Civil Aviation launched a new National Tourism Strategy 2016–2025, as well as a more concise Tourism Vision 2020 document. While the strategy has been mentioned in online media articles, a downloadable version could not be located. Tourism Vision 2020 has a brief strengths, weaknesses, opportunities, and threats analysis, and it outlines objectives. It has a one-page summary of strategy statements, present details for each statement, and lists immediate and long-term actions. It also has a map of the proposed Great Himalaya trail.

From 2010 to 2019, Nepal doubled international tourist arrivals from 603,000 to almost 1.2 million, with international tourism receipts rising from $344 million to $701 million (Table 6). Some of this growth can be attributed to the new National Tourism Strategy and Tourism Vision 2020.

Tourism Vision 2020 states: "Tourism is valued as the major contributor to a sustainable Nepal economy, having developed as an attractive, safe, exciting, and unique destination through conservation and promotion, leading to equitable distribution of tourism benefits and greater harmony in society." To realize this vision, the following objectives have been set:

- Increasing annual international tourist arrivals to 2 million by 2020.
- Increase tourism employment to 1 million.
- Foster integrated tourism infrastructure, increase tourism activities and products, generate employment in rural areas, and enhance inclusiveness of women and other minority populations.
- Develop tourism as a broad-based sector by bringing tourism into the mainstream of Nepal's socioeconomic development, supported by a coherent and enabling institutional environment.

---

92 Government of Myanmar, Ministry of Health and Tourism. COVID-19 Tourism Relief Plan. https://tourism.gov.mm/wp-content/uploads/2020/06/COVID-19-Tourism-Relief-Plan.pdf.

- Expand tourism products by enhancing community capacity to participate in tourism activities.
- Promote the image of Nepal in international tourism source markets.
- Enhance flight safety and aviation security, extend air connectivity, and improve capacity and facilities of national and international airports.
- Attract new investment in creating new tourism facilities, products, and services.

The Tourism Profile prepared by the Investment Board of Nepal, the National Tourism Strategy, and Action Plan, 2015–2024 provides direction and guidance for implementing the proposed strategy. The emphasis of Phase 1 covering 2015–2019 is on diversifying and improving services and products by expanding new tourism areas and locations to relieve congestion in certain tourism zones and developing new and improved products and services. Phase 2 covering 2020–2025 focuses on quality control, consolidation, and expansion, and strengthening Phase 1 products and targeting new high-yield markets. These objectives correspond to some of the challenges facing the country's tourism sector.

Thematic Circuit–Based Tourism in Nepal

Tourism Vision 2020 contains the development and promotion of the Great Himalaya trail as a strategy for circuit-based tourism in Nepal.

In the survey of tour operators, 62.5% of the 40 respondents said they offered circuits to international groups and 45.0% to domestic groups (Figure 23).

Some 62% of respondents were "very optimistic" about the prospects for increased international tourism and additional 23% "optimistic." For domestic tourism, some 60% said they were either "very optimistic" or "optimistic" (Figure 24).

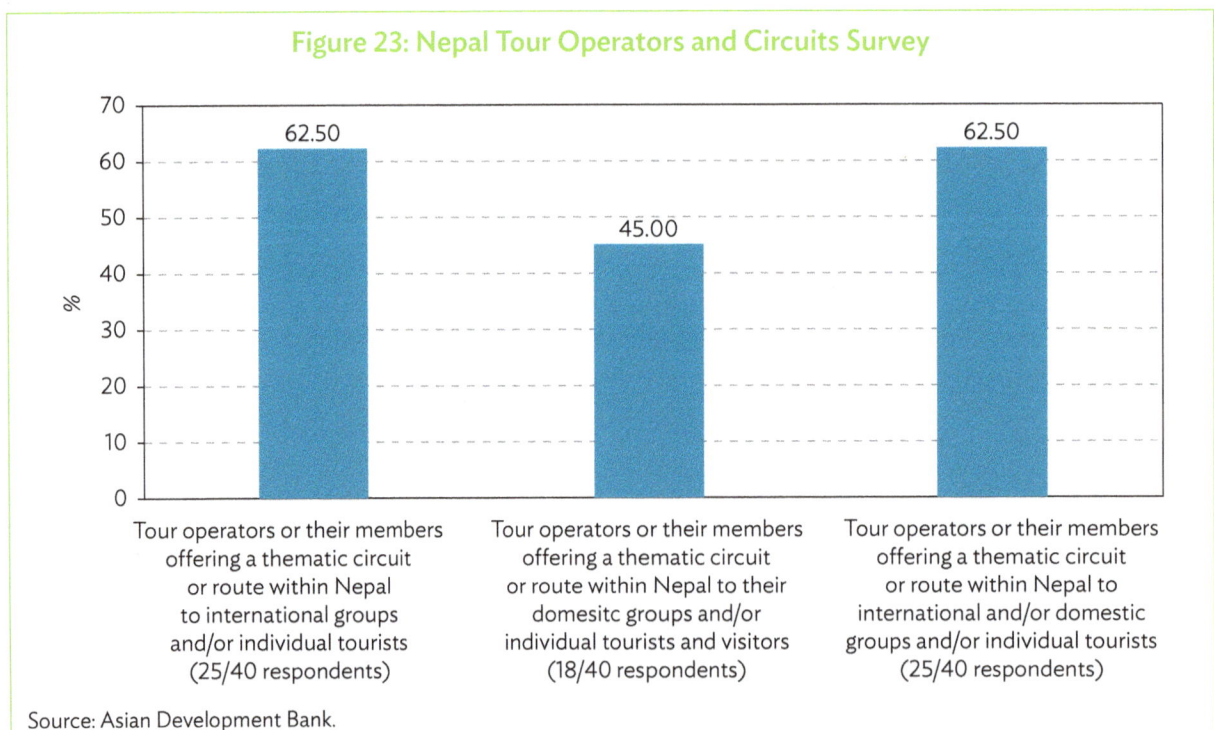

Figure 23: Nepal Tour Operators and Circuits Survey

Source: Asian Development Bank.

Figure 24: Pre-COVID-19 Optimism of Tour Operators in Nepal Survey

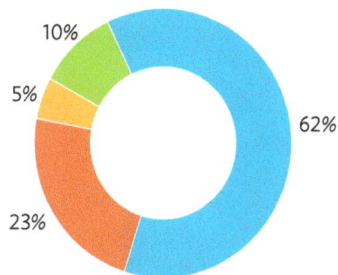

10%

5%

62%

23%

Before COVID-19, how optimistic were you or your association's members about increased international tourism to your country?

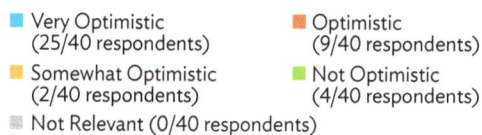

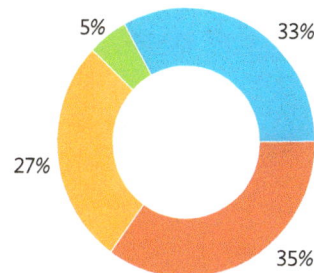

■ Very Optimistic (25/40 respondents) ■ Optimistic (9/40 respondents)
■ Somewhat Optimistic (2/40 respondents) ■ Not Optimistic (4/40 respondents)
▨ Not Relevant (0/40 respondents)

5%

33%

27%

35%

Before COVID-19, how optimistic were you or your association's members about increased domestic tourism?

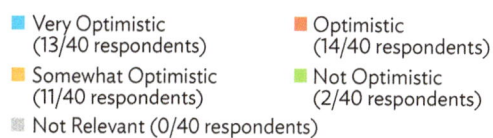

■ Very Optimistic (13/40 respondents) ■ Optimistic (14/40 respondents)
■ Somewhat Optimistic (11/40 respondents) ■ Not Optimistic (2/40 respondents)
▨ Not Relevant (0/40 respondents)

Source: Asian Development Bank.

COVID-19 Measures to Help the Tourism Industry Recover

Government assistance included the following:

- Central Bank of Nepal offering loans of up to NRs1.5 million at 2% interest rate to refinance small and cottage industries, which include tourism businesses.
- The central bank offering loans of up to NRs100 million at a 1% interest rate to refinance special and common industries, which could include some tourism businesses.[93]
- Extending loan repayments for up to 12 months.
- Supporting the expansion of domestic tourism; this has helped somewhat to ease unemployment in the industry.
- Royalty exemption for casinos during the lockdown period proposed to the Ministry of Finance.
- Mountaineering expeditions rescheduled with royalty fees postponed or withdrawn.
- Government request to property owners to provide discounts or concession on rents.
- The Nepal Tourism Board, in May 2020 made the following recommendations to the government:[94]
    - Establishing a NRs20 billion Job Retention Fund for Tourism Workforce.
    - Financial support to tourism enterprises, especially interest rate reductions and loan repayment deferments.
    - Policy interventions:
        » A mandatory Leave Travel Concession (LTC) provision for civil servants, security personnel, employees of corporations, authorities, semi-government organizations, banking sector, and corporate sectors. The leave is provided either as direct cash or an income tax rebate on the

93   Nepal24Hours.com. 2020. Nepal Central Bank Brings Refinancing Work Procedure, 2077. 2 June. https://nepal24hours.com/nepal-central-bank-brings-refinancing-work-procedure-2077.

94   Steinmetz, J. 2020. Nepal Tourism Board Government Plan to Survive COVID-19. E-turbo News. 17 May. https://eturbonews.com/572427/nepal-tourism-board-government-plan-to-survive-covid-19/.

cost of the LTC. The government estimates that 1.7 million people will use the LTC and spend NRs53 billion on domestic travel.

» Rebate on electricity and waiver on demand charges.
» Tourism promotion and infrastructure development costs will be considered corporate social responsibility expenses.
» Tax deferment of 6 months for tourism entrepreneurs.

Sri Lanka

Tourism Strategy

Currently, a new Strategic Plan for 2022-2024 is being developed in line with the development strategy of the government and SDGs in collaboration with the World Bank. Also, a Tourism Policy Document is being prepared with the assistance of UNDP and the EU.

Tourism features prominently in the country's National Policy Framework, released in December 2019, in a section titled Vistas of Prosperity and Splendor. The framework stresses an environmentally and culturally friendly tourism industry with "extensive people's participation." The framework identifies new attractions and the development of community-based tourism and ecotourism as activities—all relevant for circuit-based tourism.[95]

From 2010 to 2018, international tourist arrivals to Sri Lanka increased from 654,000 to 2.3 million but fell to 1.9 million in 2019 due to the 2019 Easter bombings. To recover from this, the Ministry of Tourism implemented the Sri Lanka Tourism Strategic Plan, 2017–2020.

The Sri Lanka Tourism Development Authority formulated the Strategic Action Plan, 2020–2022 as an interim strategy. It includes the following themes:

- A people-centric tourism sector
- Efficient public service and uplift industry standards.
- Technology-based tourism
- A safe and secure country for tourism
- A sustainable and environmentally friendly tourism sector
- New product development and promotion
- Positioning and marketing Sri Lanka
- Infrastructure and service
- Partnerships.

The vision of the Sri Lanka Tourism Strategic Action Plan, 2017-2020 is: "To be recognized as the world's finest island for memorable, authentic, and diverse experiences and to be a high-value destination offering extraordinary experiences that reflect Sri Lanka's natural and cultural heritage, are socially inclusive, environmentally responsible, and provide economic benefits to communities and the country (footnote 92)." The plan's three objectives and seven guiding principles are:

---

[95] Government of Sri Lanka. Vistas of Prosperity and Splendour. Summary of National Policy Framework. December 2019.

Objectives

- Tourism to be Sri Lanka's third-highest net foreign exchange earner, with a target of $7 billion earned in 2020
- Tourism and its supporting industries to employ 600,000 Sri Lankans, with women accounting for 10% of the workforce
- To increase daily spending per visitor to $210

Guiding Principles

- Capturing the benefits of tourism for Sri Lanka
- Democratization of economic participation
- Conservation and world-class management of assets
- Local community involvement
- Memorable experiences rooted in heritage
- Responsible destination marketing
- Safety and security for all.

Thematic Circuit-Based Tourism in Sri Lanka

P. Thilak Weerasinghe, chairperson, Lanka Sportreizen, presented several opportunities for tourism development at the October 2015 UNWTO conference. Weerasinghe described the UNESCO World Heritage Sites in Sri Lanka,—highlighting Jaya Sri Maha Bodhi, Ruwanweli Maha Seeya, Mihinthale, Temple of Tooth Relic-Kandy, Watada Geya–Polonnaruwa, and Gal Viharaya–Polonnaruwa and described the Ramayana trail that includes 55 sites, of which 20 are accessible and therefore could be the basis for a national-level thematic circuit.

In the survey of tour operators, nearly 90% of respondents said they offered circuits to international groups and/or domestic groups. Of these, 77.8% offered circuits to international groups and 48.2% offered circuits to domestic groups. Some 86% of operators said they were "very optimistic" about the prospects for increased international tourism and 7% were "optimistic." For domestic tourism, 30% were "very optimistic" and 18% "optimistic."

COVID-19 Measures to Help the Tourism Industry

The government implemented several measures to help the travel and tourism industry recover losses and restart operations. These include:

- Value-added tax exemption for the tourism industry;
- One-off payment of SLRs20,000 for tour guides registered with the Sri Lanka Tourism Development Authority (SLTDA);
- One-off payment of SLRs15,000 for tourist drivers registered with the SLTDA;
- Moratorium extension for tourism service providers from September 2020;
- Increase working capital subsidy loans from SLRs25 million to SLRs50 million;
- Registration of restaurants, tourist-friendly eating places, spa and wellness centers, spice gardens, tourist shops, and water sports centers under the SLTDA can borrow at 4% interest through the government's post COVID-19 Relief Budget via Bank of Ceylon to pay a monthly salary of SLRs15,000 per employee for six months;
- Grace period extension for vehicles leased for tourism purposes extended to 12 months without default charges;
- Support for utility payments;

- Social security scheme for individual SMEs in industry;
- Facilitated moratorium grants for tourism SMEs through international and bilateral organizations such as the EU and USAID; and
- Material grants for women entrepreneurs were made with the assistance of UNDP and Citibank.

Other measures:
- Detailed COVID-19 guidelines publicly available with the assistance of Ministry of Health;
- Appointed independent auditing firm for COVID-19 certification;
- Pandemic preparedness course for small and medium-sized enterprises;
- Annual registration (renewal) fee waived in 2020 for the accommodation sector, travel agents, tour guides, and tourist drivers; and
- Relaxed registration fees.

Thailand

Tourism Strategy

Thailand's tourism strategy is in the Second National Tourism Development Plan (NTDP) of the Ministry of Tourism and Sports strategy (2017–2021), a comprehensive document covering all issues related to tourism development. This document begins with an overview of the sector and the Tourism Vision toward 2036. The plan specifies 5-year tourism objectives and five sets of "inputs" to realize these objectives. It has five sets of 5-year "strategic axes," each with multiple measures and initiatives. The plan has a well-structured framework for development, which insists on consensus from all stakeholder groups.

From 2010 to 2019, Thailand was one of the world's fastest-growing international tourist destinations, with arrivals increasing from 15.9 million to almost 40.0 million. International tourism receipts rose from $20.1 billion to $61.5 billion in the same period. The 2019 WEF-TTCI ranked Thailand 31st out of 140 countries. Tourism, in 2019 accounted for almost 10% of GDP and 16% of employment. The country's success in tourism can, at least in part, be attributed to the implementation of its Second NTDP.[96]

The Second NTDP's vision for tourism is: "By 2036, Thailand will be a World's leading quality destination, through balanced development while leveraging "Thainess" to contribute significantly to the country's socio-economic development and wealth distribution inclusively and sustainably."[97] To realize this vision, the Second NTDP specifies the following objectives and key performance indicators, as shown in Figure 25.

To achieve these objectives and key performance indicators, the plan includes five sets of 5-year "strategic axes," each with multiple measures and initiatives. The axes are depicted in Figure 26 and cover a comprehensive range of tourism development and marketing issues and subjects.

Of the five strategic axes, Strategy 5 is the most directly relevant for thematic circuits with other countries. Strategy 5 led to the establishment of Tourism Intelligence Center, and it includes measure 5.4, which aims to improve international collaboration for developing tourism and specifies the following four initiatives:

---

[96]   Government of Thailand, Ministry of Tourism and Sport. 2017. *Second National Tourism Development Plan*. Bangkok.
[97]   UNWTO. 2017. *Measuring Sustainable Tourism*. Madrid.

Figure 25: Thailand Second National Tourism Development Plan, 2017–2021

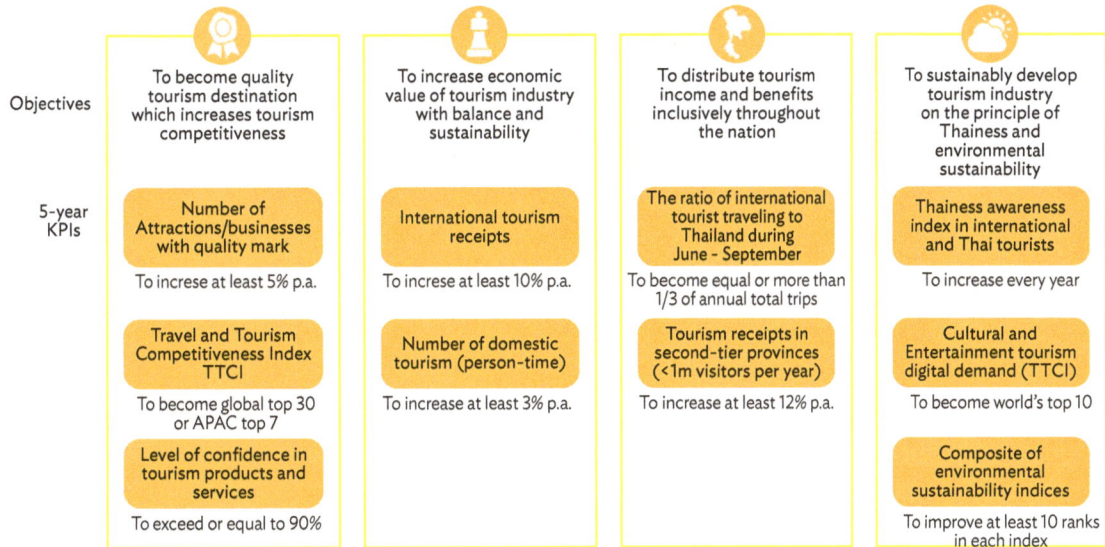

Objectives	To become quality tourism destination which increases tourism competitiveness	To increase economic value of tourism industry with balance and sustainability	To distribute tourism income and benefits inclusively throughout the nation	To sustainably develop tourism industry on the principle of Thainess and environmental sustainability
5-year KPIs	Number of Attractions/businesses with quality mark — To increse at least 5% p.a.	International tourism receipts — To increse at least 10% p.a.	The ratio of international tourist traveling to Thailand during June - September — To become equal or more than 1/3 of annual total trips	Thainess awareness index in international and Thai tourists — To increase every year
	Travel and Tourism Competitiveness Index TTCI — To become global top 30 or APAC top 7	Number of domestic tourism (person-time) — To increase at least 3% p.a.	Tourism receipts in second-tier provinces (<1m visitors per year) — To increase at least 12% p.a.	Cultural and Entertainment tourism digital demand (TTCI) — To become world's top 10
	Level of confidence in tourism products and services — To exceed or equal to 90%			Composite of environmental sustainability indices — To improve at least 10 ranks in each index

KPI = key performance indicator, p.a. = per annum.

Source: Government of Thailand, Ministry of Tourism and Sport. 2017. *Second National Tourism Development Plan*. Bangkok. p. 18.

Figure 26: Five-Year Strategic Axes of Thailand Second National Tourism Development Plan, 2017–2021

Strategy 1 — Development of tourism attractions, products and services including the encouragement of sustainability, environmental friendly, and Thainess integrity of attractions

Strategy 2 — Development and improvement of supporting infrastructure and amenities without inflicting negative impact to the local communities and environment

Strategy 3 — Development of tourism human capital's potential and the development of tourism consciousness among Thai citizens

Strategy 4 — Creation of balance between tourst target groups through targeted marketing that embraces Thainess and creation of confidence among tourists

Strategy 5 — Organization of collaboration and integration among public sectors, private sectors and general public in tourism development and management including international cooperation

KPI = key performance indicator.

Source: Government of Thailand, Ministry of Tourism and Sport. 2017. *Second National Tourism Development Plan*. Bangkok. P. 18.

Initiative 5.4.1. Promote regional tourism corridors among such as Kong River tourism routes. One of the recommended actions of this initiative is the identification of potential tourism routes focusing on Cambodia, the Lao PDR, Myanmar, Viet Nam, and Thailand routes first. Under this initiative: "Thailand should look into establishing these routes as official tourism corridors. This can be achieved by outlining tourism corridors' offerings from a tourism lifecycle perspective and designing marketing, branding, and promotion strategies."

Initiative 5.4.2: Promote joint marketing efforts with regional government partners, such as co-developing a regional tourism strategy and sharing data to boost marketing effectiveness.

Initiative 5.4.3: Enhance connectivity in regional tourism and support the development of regional tourism corridors.

Initiative 5.4.4: Making regional traveling easier, reducing administrative difficulties of entry, and creating a single unified regional visa.

The potential for thematic circuits between Thailand and other countries was highlighted in 2015, when Sriporn Bhekanandana, deputy executive director, ASEAN, South Asia, and South Pacific Region, at that time, presented several opportunities for tourism development in Thailand at the UNWTO conference. Bhekanandana highlighted opportunities for regional tourism in BIMSTEC countries that could leverage the benefits of regional connectivity routes, particularly the India–Myanmar–Thailand Trilateral Highway, which connects South Asia with Southeast Asia linking Moreh (India) with Mae Sot, Tak Province (Thailand).

Following up on this highway prospect, Bhekanandana explained that the Tourism Authority of Thailand, the Royal Bhutanese Embassy, and the Thai Rung Union Car PLC[98] organized the Bhutan–Thailand Friendship Caravan in 21–28 June 2019, which covered a 3,000 km journey that followed the India–Myanmar–Thailand Trilateral Highway into Bhutan. The caravan took place under the Two Kingdoms One Destination concept to celebrate the 30th anniversary of diplomatic relations between Thailand and Bhutan. The hope was that this initiative would set the stage for boosting regional connectivity and economic growth, especially among people.[99]

Projects like the Trilateral Highway and the Bhutan–Thailand caravan (which aims to stimulate travel on the highway) could strengthen circuits for sustainable tourism to and from Thailand, as well as for developing sustainable regional tourism.

Bhekanandana drew attention to existing Buddhist trails that could be explored and developed in the Bay of Bengal region and also highlighted tourist attractions in Thailand based on Buddhist heritage, including the possibility of visiting nine "merit" temples in a day.[100]

The survey conducted by this report with tour operators in BIMSTEC countries received responses from only six in Thailand. Three of them said they offered a thematic circuit or route within Thailand to international groups and/ or individual tourists; only one offered circuits to domestic groups. Nearly all were "very optimistic" or "optimistic" about the prospects of increased international tourism to and domestic tourism within Thailand.

---

98   Thai Rung Union Car PLC designs and manufactures automotive parts.
99   Notes from a telephone interview conducted by ADB with Sriporn Bhekanandana, mid-August 2020 (specific day not reported).
100   Report of the First Meeting of BIMSTEC Network of Tour Operators. 7 July 2017. Delhi.

As of 22 August 2021, the COVID-19 Travel Regulations Map of the International Air Transport Association (IATA) showed that Thailand was closed to passengers and air crews from some countries until 30 September 2021.[101] This did not apply, however, to Thai residents, spouses of Thai nationals, and other exempted categories. And notably, those traveling with an Asia-Pacific Economic Cooperation Business Travel Card were also exempt, as were nationals from 55 countries. Because of these restrictions, domestic tourism had become even more important to Thailand.

COVID-19 Measures to Help the Tourism Industry

The government has implemented several measures to help the travel and tourism industry through the COVID-19 pandemic. An important feature of these measures is that they are holistic in that they have sought to help the industry rather than individual businesses, as the following measures show:

- Thailand Safety and Health Administration provided guidelines for all travel-related businesses to enhance their health and safety standards, and the quality of their products and services.
- Drafting a regulation to decrease the business licensing application fee for travel and tourism businesses.
- Tourism-related government agencies were drafting, at the time of writing, budget allocation requests for projects that aimed to help rebuild the tourism sector.
- Funding had been requested for the following projects:
  - Building confidence for Thai people and tourists.
  - Promoting a good image of Thailand.
  - Stimulating domestic tourism.
  - Developing and analyzing a big data storage system.
  - Developing the capacity of tourism personnel.
  - Increasing tourist safety through CCTV installation in Chiang Mai, Surat Thani, Krabi, Chonburi, and Phuket provinces.
  - Developing cybersecurity systems for crime prevention.
  - Building community capacity for sustainable tourism.[102]

To explain entry procedures, the Ministry of Tourism and Sports by Tourism Authority of Thailand produced a comprehensive video that explained the process for travelers entering the country (Figure 27).

Figure 27: Thailand Ministry of Tourism and Sports Entry Process Video

Source: Tourism Authority of Thailand presented via Vimeo at https://vimeo.com/491541327.

---

101 IATA. COVID-19 Travel Regulations Map. https://www.iatatravelcentre.com/world.php.
102 UNWTO. 2020, *COVID-19: Measures to Support the Travel and Tourism Sector*. 2 April. https://www.unwto.org/.

Illustrative Examples from Member Countries for an Updated Plan of Action for Tourism Development and Promotion for the BIMSTEC Region

The following are illustrative examples taken from country profiles that were developed for an earlier iteration of this report, which could be useful for exchange among BIMSTEC countries via the Secretariat for updating the action plan and developing cross-border circuits:

COVID-19 Measures

- BIMSTEC states are pursuing national programs to reduce the spread of COVID-19 and assisting in the recovery of their tourism sectors. These programs span a range of actions including fiscal and monetary policy, employment assistance, public–private partnerships, health and safety, and domestic tourism. The UNWTO's Dashboard for COVID-19 Measures to Support Travel and Tourism provides a comprehensive aggregation of measures for most countries, including all BIMSTEC countries. Establishing and maintaining a BIMSTEC-specific dashboard via a central information repository could include the UNWTO data and specific examples from each country.

Infrastructure

- BIMSTEC states are pursuing national infrastructure development plans, especially for ground, maritime, and air connectivity for cross-border connectivity, some of which were with donor support. Road improvements in Bangladesh will improve access to Buddhist sites, such as Vihara Paharpar, and increase the possibility of establishing new Buddhist and other thematic circuits. Some 56 transport connectivity projects are listed in Annex 7 (airport projects), Annex 8 (road and rail projects), and Annex 9 (maritime projects), several of which are benefiting or could benefit connecting tourist destinations and attractions across BIMSTEC.
- An important point worth mentioning again is that Bhutan's infrastructure development is driven by its guiding principle of "high value, low volume" tourism, which strengthens the sustainability of its tourism sector.
- Thailand's Second NTDP is notable for emphasizing the "improvement of supporting infrastructure and amenities without inflicting negative impact to the local communities and environment." This concept is worth replicating in other BIMSTEC countries. Thailand's infrastructure development is being boosted by initial work on the Trilateral Highway into Bhutan and the NTDP's Strategy 5, particularly the following initiatives:
  - Initiative 5.4.1: Promoting regional tourism corridors among "peers," such as Kong River tourism routes.
  - Initiative 5.4.3: Enhancing connectivity in regional tourism and supporting the development of regional tourism corridors.
- Sri Lanka is pursuing multiple aviation and maritime port improvement and expansion plans for better domestic and international connectivity.
- India's Ministry of Tourism cites "quality tourism infrastructure" as a function of the Ministry and notes that 40% of its expenditure goes on infrastructure-focused tourism schemes, such as Swadesh Darshan and the National Mission on Pilgrimage Rejuvenation and Spiritual Heritage Augmentation Drive. It gives funding support to state governments for implementation and administration of at least 13 circuits. Circuit development that stresses improved visitor-related infrastructure, including increasing the number of hotel rooms, is central to the ministry's programs. These are the most developed for circuits among BIMSTEC countries.

Product and Marketing

- Most of the tour operators that took part in the survey said they either already offered or would offer thematic circuits if available. They welcomed government-led improvements in infrastructure that improve access to and the protection and preservation of sites and attractions. These measures, in turn, strengthen the quality and marketability of these sites, thereby also strengthening potential returns on investments and benefits for local communities.
- The diversity of ecotourism and cultural tourism offers are a BIMSTEC-wide product strength that appeals to a wide cross-section of domestic and international visitors. These offers are a strong basis for the further development of circuits within each country. Progress in developing cross-border circuits will depend on the lifting of COVID-19 travel restrictions in BIMSTEC countries—an uncertain prospect at the time of writing. In the meantime, internal circuits can be developed for domestic markets.
- World Heritage sites, such as the Sundarbans tiger reserves in Bangladesh, the ancient city of Polonnaruwa in Sri Lanka, and Lumbini, the birthplace of the Buddha in Nepal, offer strong anchors for existing and potential circuits.
- Cross-border river cruises, such as one proposed by the Indian luxury travel operator Exotic Heritage Group, have potential for river-based international circuits.
- The Tourism Council of Bhutan is marketing the country as a filming destination, a strategy that has already proven effective; this can also boost international marketing. Bhutan has considerable experience in this area, with IMDb listing 67 films having been made there. This is an effective strategy to emphasize for other BIMSTEC countries individually and as a region.[103]
- India, Myanmar, and Thailand were increasingly marketing to domestic markets because of the collapse in international travel caused by the COVID-19 pandemic since international travel was expected to be slow into 2022.
- Sri Lanka is cooperating with India to promote tourism, including operating a circuit between the two countries, a cruise service between Kochi and Kerala, and special fares on Sri Lankan Airlines from Colombo to Kerala.
- Thailand's National Tourism Development Plan emphasizes "Thainess and [the] creation of confidence among tourists." It also emphasizes the development of tourism that reinforces cultural and environmental sustainability and supports balanced development. These are all elements that may be unique and worth replicating across BIMSTEC.
- The WEF-TTCI ranked Thailand 18th in effectiveness of marketing, 4th in natural tourism digital demand, and 22nd in cultural and entertainment.

Human Resources

- Each BIMSTEC member state has government-sponsored and/or private hospitality and tourism training institutes, such as the Royal Institute for Tourism and Hospitality in Bhutan and the 41 Institutes of Hotel Management in India, which helped provide some of the trained staff (pre-COVID-19) needed for the industry in each country.
- Sri Lanka's Tourism Strategic Plan, 2017-2020 lays out plans to prioritize human resources, including greater participation for women and local communities.
- Sri Lanka also works with Kerala Tourism in India on human resource development and skills training.
- A notable feature of Myanmar's Thirty-Year Long-Term Education Development Plan, 2001–2030 is the target of training more than 563,000 people in tourism. The plan includes the country's first 4-year bachelor's degree in tourism and a postgraduate diploma in tourism.

---

103   For a list of films with Bhutan as location, see https://www.imdb.com/search/title/?locations=Bhutan.

- Thailand's National Tourism Development Plan focuses on the "development of tourism consciousness among Thai citizens," again a human resources element that is possibly unique among BIMSTEC members. Because Thailand ranks a strong 27th on human resources/labor market and 26th on qualification of the labor force in the 2019 WEF-TTCI, the country's approach to developing human resources in the travel and tourism industry merits deeper analysis by other BIMSTEC countries—an example the BIMSTEC Secretariat should encourage.

Governance, Policy, and Investment

- Most BIMSTEC countries identify tourism as a sector for investment, albeit with varying degrees of emphasis. Bangladesh, for example, offers tax incentives in the form of tax holidays, and the WEF-TTCI ranks the country's labor and tax contributions rates as the most competitive in the world.
- It is important that BIMSTEC countries update their tourism policies to ensure quality and sustainability. Bhutan's tourism policy has long been driven by the principle of "high value, low volume," which has ensured high quality and sustainability. While the principle might not be completely replicable in every country, the recognition that there are limits to growth and limiting visitors can increase value and quality is more important than ever as social distancing becomes enshrined in destination management post-pandemic.
- Nepal's tourism policy helpfully includes licensing requirements for travel and trekking agencies, hotels, restaurants, and bars; requirements and restrictions for climbing Himalayan peaks; tour guide requirements; and requirements for starting tourism businesses.
- Myanmar's tourism policies and the Tourism Master Plan provide a comprehensive basis on which to improve and sustainably develop tourism. The plan lays out a well-organized framework of guiding principles and strategic programs, which specify activities to address infrastructure, product and services, marketing and branding, human resources, and destination management.
- Sri Lanka's Tourism Strategic Plan, 2017-2020 is a well-organized document that covers a range of issue areas with a "transformational theme" to improve institutional performance, governance, and regulations. An entire chapter is devoted to improving governance and regulation. This could be a useful guide for other BIMSTEC countries.

Proposed Vision Statement for BIMSTEC Tourism

The need is urgent to plan for a more strategic approach to tackle the challenges for the sustainable development of tourism in the BIMSTEC region. These challenges include infrastructure, product and marketing, human resources, policy issues, and investment. While COVID-19 magnifies these challenges, tourism can be a fast-track path for economic recovery and for increased sustainability, inclusiveness, and economic integration in the region. The proposed vision and mission statements for developing BIMSTEC as a regional tourism destination are:

- **Vision.** By 2030, the BIMSTEC region becomes a global center for experiential circuit-based tours, which benefits each member country through increased visitor spending, business, and investment, maximizing benefits and synergies among members and contributing to the achievement of the UN Sustainable Development Goals.
- **Mission.** The BIMSTEC Secretariat coordinates public and private sector stakeholders in developing and offering unique circuit-based tours that showcase the vast diversity in the region's national, cultural, religious, and nature-based heritage, as well as promoting the equitable distribution of tourism's social and economic benefits to society.

Sustainability Principles: Key for Realizing the Vision and Achieving the Mission

Because of the global impact of COVID-19 on tourism, the twin issues of sustainable tourism and a resilient tourism industry are more important than ever. Discussions with stakeholders in BIMSTEC countries for this report showed that sustainable tourism is a high-profile issue as the pandemic has resulted in raised standards, especially in health and safety, that are being demanded by operators and their customers. A government official in India mentioned that, because of COVID-19, he expects increased demand for responsible and "socially distant" nature-based tourism.

The UNWTO's definition of and principles for sustainable tourism and the SDGs provide a firm basis for developing BIMSTEC tourism. The UNWTO defines sustainable tourism as: "tourism that takes full account of its current and future economic, social, and environmental impacts, addressing the needs of visitors, the industry, the environment, and host communities."[104] The UNWTO has formulated guidelines and management practices applicable to all forms of tourism in all types of destinations and niche tourism segments, including themed circuits. Its sustainability principles are based on the so-called triple bottom line that considers the environmental, economic, and sociocultural aspects of sustainable tourism development.

The SDGs provide guidance for sustainable tourism in destinations around the world and could do the same for developing tourism circuits in BIMSTEC countries. Figure 28 lists the 17 SDGs and Annex 11 describes how each relates to tourism. The SDGs that are most relevant to developing circuits in the BIMSTEC region are indicated for each of the strategic objectives.

BIMSTEC governments are striving to uphold the principles of sustainability, and how they achieve it can inform each other's national strategies and plans. And as noted later in the recommended actions, the Secretariat can play a regional coordinating role to facilitate an exchange of successes and best practices, including examples from the private sector.

Figure 28: Sustainable Development Goals

Source: United Nations.

---

[104] UNWTO. 2017. *Measuring Sustainable Tourism*. Madrid.

Strategic Objectives and Key Action Areas

Realizing the proposed BIMSTEC tourism vision and achieving the Secretariat's mission requires building on and, where possible, developing new initiatives in the context of the issue areas just discussed. National and cross-border development of thematic circuits can catalyze sustainable tourism in BIMSTEC countries.

The strategic objectives and recommended action areas can address the challenges BIMSTEC member states face in developing their travel and tourism industries, and implementation of the action areas can help build on their strengths, resilience, and opportunities for developing circuits. Several action areas involve setting up and populating an online repository and app for sharing knowledge and experiences among BIMSTEC member states and other organizations that already do this, like ASEAN and the GMS, as well as facilitating the exchange of best practices for developing national and cross-border tourism circuits.

An online repository and app could be an extension of the BIMSTEC Tourism Information Center once it is set up. The center was proposed in the Kathmandu Declaration on Tourism Cooperation on 29 August 2006. The center could be entirely virtual and maintained on Secretariat servers. It should be easy for member states' tourism authorities to upload and download material, and for the industry and prospective visitors to register interest in and download information on circuits and other tourism information.

All the Action Plan recommendations are still relevant, although some modifications for market and technology developments are needed since it was formulated back in 2006.

Strategic Objective 1: Control and mitigate COVID-19 impacts on tourism

Addressing issue area 1 requires government action and leadership to control COVID-19 and implement health and safety measures in partnership with the private sector. Regional communication and cooperation are critical to these actions.

Recommended actions

Every BIMSTEC member needs to maximize coordination of implementation protocols and measures to control and recover from the spread of COVID-19, thereby adopting best practices and learning lessons. These measures include travel restrictions, improved health and safety measures, and the several policy areas just mentioned, which are being monitored through the UNWTO Dashboard Policy Measures Tracker (Figure 29).

Exchanging best practices and policies for controlling the spread of COVID-19 are important for eventually restarting cross-border travel through circuits. To the extent possible, it would be helpful to harmonize COVID-19 mitigation and health guidelines and standards across the tourism value chains of BIMSTEC countries. The Secretariat could assist by disseminating this information and exchanging best practices for raising awareness of the measures and protocols via a virtual BIMSTEC Information Center.

While Thailand and some other BIMSTEC countries have comprehensive COVID-19 programs, it would be helpful to encourage the adoption of BIMSTEC-wide healthy destination and property designations for cross-border travel, such as via circuits, so that visitors, tour operators, and prospective investors are assured of finding similar COVID-19 health measures from country to country. The WTTC's Safe Travels protocols and stamp have been adopted by multiple public and private tourism organizations, including the Sri Lanka Tourism Development Authority, the Goa Department of Tourism, and five tourism industry associations in India and Thailand. Alternatively, programs

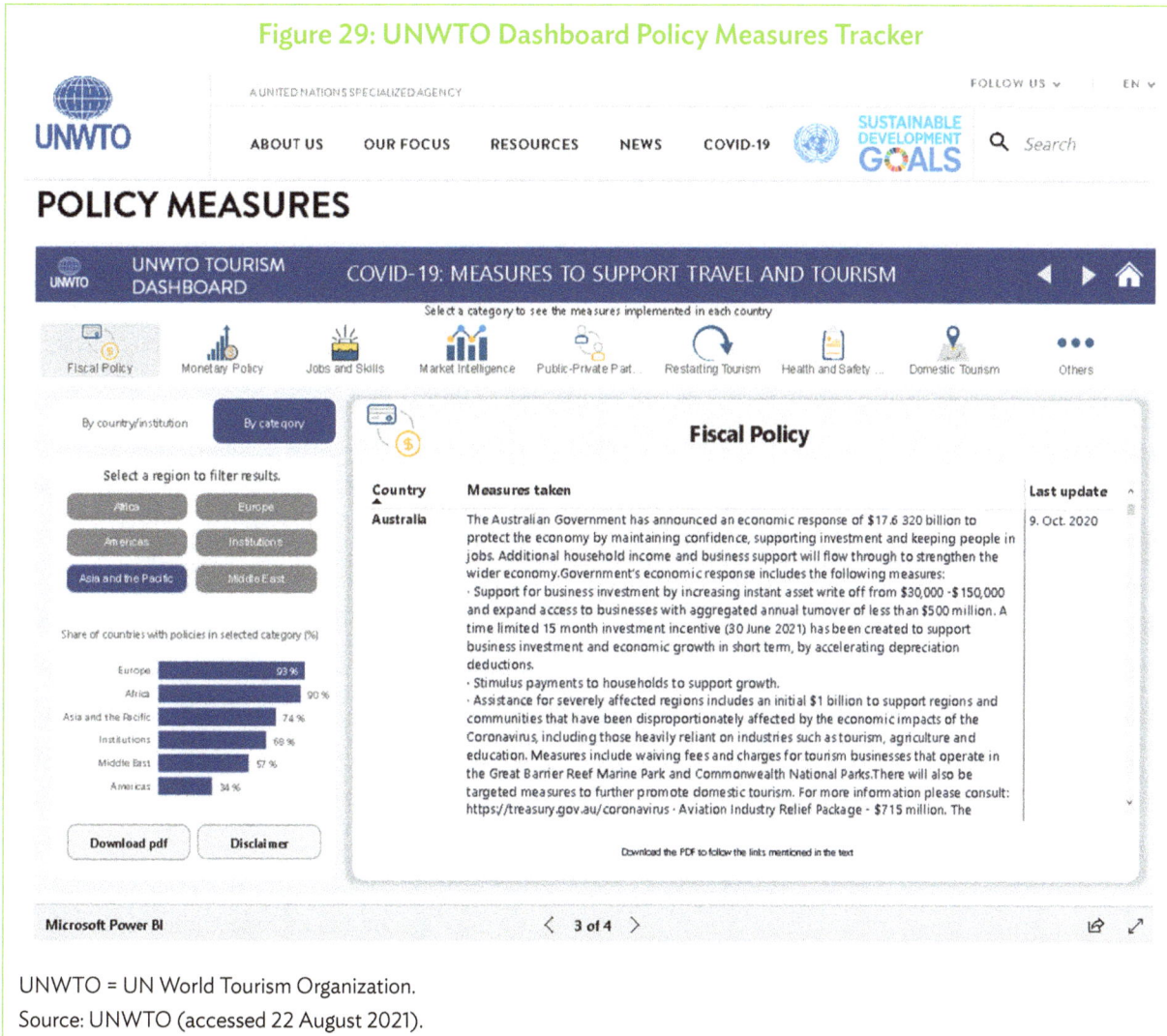

Figure 29: UNWTO Dashboard Policy Measures Tracker

UNWTO = UN World Tourism Organization.
Source: UNWTO (accessed 22 August 2021).

could be promoted that are compliant with international standards and programs, such as those of the WTTC. Indeed, applying these standards is a WTTC registration requirement.

Access to the action plan can be virtual and the website can serve as a BIMSTEC portal.

These action areas could contribute to achieving the following SDGs:

Strategic Objective 2: Recover lost tourism jobs and businesses

Addressing issue area 2 requires immediate government support across the travel and tourism value chain to help businesses recover from the economic impact of the COVID-19 pandemic and workers who lost jobs or were furloughed to get back to work.

Recommended actions

2.1   Assess the situation in each BIMSTEC country through the joint efforts of government tourism organizations and private sector associations. Unofficial results from the survey conducted by ADB workshop on 25 November 2020 show that the situation was dire, with thousands of tourism jobs lost in each BIMSTEC country (Table 20). More precise data are needed to be able to plan better for supporting the industry and assisting in reemployment.

2.2   Some tourism businesses in BIMSTEC countries, including hotels and restaurants, have been able to continue with reduced capacity but still below pre-COVID-19 levels. A partial solution, which is being encouraged and promoted by BIMSTEC governments, is to focus on serving mostly domestic visitors and limited numbers of international visitors. Public–private partnerships for domestic marketing have been helpful, as has been done in Thailand, India, and some other BIMSTEC countries.

2.3   Government assistance to businesses via subsidized low-interest loans, training support, temporary employment for furloughed workers, and financing for enhanced health and safety measures are all initiatives that could be expanded. Even though COVID-19 vaccination programs are progressing, they will take time to complete and for the industry to recover.

2.4   Connection to 2006 BIMSTEC Tourism Action Plan: Many lessons have been learned over the past year on effective measures for mitigating the COVID-19 crisis. Expanding health and safety measures has been an especially important lesson for BIMSTEC members and their tourism sectors. The effect of the financial measures taken to mitigate the impact of the crisis are uncertain for that very reason that it is not known how long the crisis will last—an important point that government's need to consider.

These action areas could contribute to achieving the following SDGs:

Strategic Objective 3: Improve tourism-related infrastructure

Addressing issue areas 3 and 4 covers a broad range of needs for improving tourism and transport infrastructure, and connectivity specific to the further development of cross-border thematic circuits. Sites along existing and potential circuits should be the focus of infrastructure improvements.

Recommended actions

All BIMSTEC countries are pursuing plans and projects to improve tourism infrastructure and, in the process, adopting best practices and learning lessons.

3.1   Establish a central online BIMSTEC repository and application (app) of plans and progress reports about tourism-related infrastructure projects to help inform members about best practices and lessons learned in developing and maintaining visitor infrastructure. The repository and app could be set up so that each member could easily submit descriptions online. This could be equivalent to the BIMSTEC Tourism Information Center proposed in the Kathmandu Declaration on Tourism Cooperation on 29 August 2006, but it would extend beyond the original visitor information function to include project opportunities, particularly along circuits. The focal point for this initiative would need to be decided.

3.2   BIMSTEC should encourage a focus on opportunities for PPPs for financing, developing, and maintaining infrastructure projects. Examples could include the following:

- **Roadside amenity complexes.** These are needed to fill the infrastructure gap in tourist services. Services could include well-maintained toilets and small business opportunities, such as petrol stations, restaurants, cafes, and shops for snacks and souvenirs. The government could provide the land, utilities, and waste management to investors who set up and run these complexes. The goal would be to establish these complexes along circuit roads, and they could become a replicable model for a chain of amenity complexes at other tourist destinations.
- **Road examples for one or more complexes.** These could be the roads along the Buddhist circuit, which connect Bodh Gaya and Patna or Rajir to Patna, and the India–Myanmar–Thailand Trilateral Highway. Along the 1,380 km highway are multiple possible sites of interest to visitors, including the temples in and around Mandalay and the important Buddhist pilgrimage site of Kyaiktiyo (Myanmar). In December 2020, Bangladesh asked to connect to the Highway from West Bengal (Hilli) to Meghalaya (Mahendraganj).
- Hotels are needed at points along circuits, and this will provide opportunities for PPPs whereby the government provides land, utilities, and waste management and investors develop and operate hotels and related services, such as restaurants and cafes. The BIMSTEC Secretariat should encourage governments to work with their private associations to identify needs at specific sites along existing and potential circuits.

3.3   Where possible, the BIMSTEC Secretariat will encourage the development of infrastructure in line with SDG 12 for sustainable consumption and production patterns. Examples across BIMSTEC countries include Bhutan's guiding principle of "high value, low volume," and Thailand's focus on improving supporting infrastructure and amenities without negative impacts on local communities and the environment, as well as their emphasis on infrastructure improvements that expand inclusiveness and access for the disabled and elderly.[105]

3.4   Focus on infrastructure projects to strengthen internal circuits and cross-border projects that could boost regional circuits. Examples of both internal and cross-border circuits include:

---

[105]   Government of Thailand, Ministry of Tourism and Sport. 2017. Second National Tourism Development Plan. Bangkok.

- Ramayana circuit for Nepal and India[106]
- Buddhist circuit for Nepal and India
- Himalaya circuit between Nepal and Bhutan
- River cruise circuit between Bangladesh and India
- Ocean cruise circuit between Sri Lanka and Kerala, India
- Heritage site circuit from Thailand to Myanmar, starting from Viet Nam
- India's Swadesh Darshan circuit program.

3.5   BIMSTEC should encourage government support for infrastructure improvements along regional circuits, as this helps improve connectivity and the quality of visitor experiences at destinations along the circuit. Examples, including those in the GMS Tourism Strategy, could include:

- Standardized tourism signage across BIMSTEC countries based on uniform standards agreed by members;
- Trees and landscaping features that highlight ecological diversity that, in turn, become natural attractions for visitors;
- Lighting for paths, parking, and sites; and
- Rest stops with toilets, tourist information, shops, and restaurants, perhaps smaller versions of roadside amenity complexes.

3.6   BIMSTEC states could benefit from India's Swadesh Darshan Scheme thematic circuit toolkit in their efforts to establish or adapt an infrastructure framework. The Swadesh Darshan toolkit provides a framework for evaluating circuit potential and perhaps establishing and maintaining circuits. At the same time, the reasons for the scheme's lack of progress requires further evaluation to provide learning lessons.

3.7   Another important tourism-related infrastructure area recommendation is the improvement of health and hygiene facilities. As indicated by the 2019 WEF-TTCI rankings, most BIMSTEC members would benefit from donor assistance to improve measures and facilities for tourism-related issues to health and hygiene, physician density, basic sanitation, and drinking water (Table 21). It is not within the scope of this report to discuss possible solutions for all these areas, deferring instead to Strategic Objective 1 where measures to control and mitigate COVID-19 are recommended. Implementing these measures will address some, if not all, of these needs.

3.8   Connection to the Plan of Action for Tourism Development and Promotion for the BIMSTEC Region: Implementing the actions in points 3.6–3.8 could help achieve the plan's recommendations for tour packages and transport access facilitated by donor assistance (another plan recommendation).

3.9   BIMSTEC is facilitating the restart of low-cost carrier operations. BIMSTEC could help advance this by conducting a route market assessment to prioritize the restart of selected route pairings and include these in revised tourism marketing and promotion campaigns. The rapid growth of this aviation segment helped increase intraregional tourism before COVID-19 forced many of these carriers into bankruptcy. Getting the low-cost carrier market up and running again will be important for tourism to recover to its pre-COVID-19 trend and for stimulating the launch of new circuits.

---

[106]   A direct bus service was established in 2018 between Janakpur in Nepal, the birthplace of the Hindu Goddess Sita to Ayodhya in Uttar Pradesh, India.

3.10 BIMSTEC is assisting in coordinating the implementation of the 2021 BIMSTEC Transport Connectivity Master Plan. The plan is expected to boost cross-border linkages and circuit road connectivity across many road projects. BIMSTEC's role in a tourism context could be to facilitate coordination between tourism and transport stakeholders to expedite projects that could boost the enhancement or development of circuits in each country and across borders.

3.11 BIMSTEC should facilitate the implementation of a BIMSTEC motor vehicle agreement, which was being negotiated as of October 2020. Details of the agreement were unavailable at the time of writing, but if it is similar to the agreement signed on 15 June 2015 by the transport ministers of Bangladesh, Bhutan, India, and Nepal, it would provide a framework for the uninterrupted flow of motor vehicle traffic between and among BIMSTEC countries. Implementing this agreement could provide a multilateral framework for the movement of tourist vehicles across borders and so boost intraregional visits to single and multi-country circuits. This would, for example, allow tour operators from Nepal to take groups across the Indian border to continue Buddhist circuit tours. The BIMSTEC motor vehicle agreement, when it is final, should also be reviewed to allow tourist vehicles to cross borders with a Carnet de Passages en Douane access. For now, it appears that not all BIMSTEC countries will accept or require this access, but this should nevertheless be examined further. It would also be helpful to have mutual recognition of national driving licenses, which is part of the SAARC tourism plan.

3.12 Connection to the Plan of Action for Tourism Development and Promotion for the BIMSTEC Region: The plan recommended improved air, land, and sea accessibility, all of which could be addressed via the ADB BIMSTEC Transport Connectivity Master Plan.

These action areas could contribute to achieving the following SDGs:

Strategic Objective 4: Stimulate sustainable product development and marketing of circuits

This strategic objective addresses issue area 5, which focuses on solutions to marketing and product needs and challenges. The objective is intended to help stimulate the sustainable product development of circuits along with corresponding marketing and branding. The UNWTO's definition of sustainable tourism and the SDGs should be guideposts for these recommendations.

Recommended actions

Although the BIMSTEC region is rich in cultural and nature-based tourism assets, attractions, and activities, it is not recognized as a single regionally integrated tourist destination by visitors or tour operators. This could be remedied by BIMSTEC-led coordination with member countries on activities related to joint branding, market research, and marketing.

4.1 Conduct an inventory of existing assets that could comprise circuits, which could then form the base materials for the online repository and app of the BIMSTEC tourism information repository. It is recommended that the repository includes:

- A function that would allow tourism administration officials and tour operators to easily upload information, blog postings, videos, and images.
- As with the ASEAN tourism portal, tour operator itineraries could also be uploaded of circuit tours, especially those that include World Heritage sites and important cultural and natural heritage sites in each country. World Heritage sites are particularly noteworthy since they offer prime examples of sustainable destination management.
- Inclusion of river and ocean cruises in the repository and app.

Consider developing this in collaboration with or linking it to the Pacific Asia Travel Association, which is already well connected with tour operators throughout the region.

4.2   Include lists of BIMSTEC tour operators on the repository from each member country who are either offering circuits or are interested in doing so in cooperation with national tourism offices. A list of tour operators in each BIMSTEC country was proposed in the 2006 BIMSTEC Tourism Action Plan. This should be developed in close cooperation with the private tourism associations in each BIMSTEC country. Of the over 200 respondents to this report's survey of tour operators, at least half said they offered circuits. In addition to helping to market each country's internal and, when possible, multi-country circuits, these operators can also be an important source of feedback for improvements to each site along the circuits. They could also be involved in supporting familiarization tours for circuits when travel COVID-19 restrictions are eased or lifted. These actions are extensions of the tour package and familiarization trip actions in the 2006 action plan.

4.3   Exchange domestic marketing strategies via the repository and online roundtables offered by the Secretariat. Domestic tourism has to some extent helped fill the vacuum caused by the collapse in international tourism caused COVID-19, particularly in India, Myanmar, and Thailand. Because this trend is expected to continue once the pandemic passes, lessons learned by each BIMSTEC member on domestic tourism could be helpful in the exchange of domestic marketing strategies.

4.4   Exchange best practices and lessons learned for fostering confidence and awareness among tourists and the nationals of BIMSTEC countries on tourism that supports cultural and nature-based sustainability. These elements will strengthen each country's circuit offers since visitors are introduced to culture and nature-based experiences, and this will also help local communities to engage in these circuits.

4.5   BIMSTEC should explore setting up a dashboard similar to the ASEAN Visitor Arrivals Dashboard. As countries emerge from COVID-19 and begin to ramp up marketing, accurate data on tourism-generating markets will be more important than ever.

4.6   Reactivate the BIMSTEC Tourism Working Group to oversee the development of a BIMSTEC tourism strategy and brand using portal information and market research. The working group should include public and private participants.

4.7   BIMSTEC leads members in the development of a regional tourism strategy that could include the following:

- The BIMSTEC vision, mission, and brand enable the region to become a global circuits leader.
- An outreach strategy for BIMSTEC national tourism organizations to connect with film location scouts and production companies so that possible tourism circuit sites could be promoted as

filming locations. Often called film-induced tourism, this helps destination marketing. For example, the filming of the *Harry Potter* series increased tourism by 50% to several locations in the United Kingdom. The Tourism Council of Bhutan has a deliberate strategy of attracting film productions, which could inform the approaches of other BIMSTEC countries and perhaps lead to a BIMSTEC-wide regional strategy on this, since productions often involve more than one country.[107]

- Sample tour packages from tour operators emphasizing regional circuits that also include tours for ecotourism and adventure travel, as well as other segments, as specified by national public and private stakeholders. These were included in the original 2006 BIMSTEC Tourism Action Plan.
- Product development strategies including the following:
  - Wellness tourism as a regional offer—this was one of the fastest-growing segments pre-COVID-19. A wellness-based circuit would be highly attractive to investors, tour operators, and visitors (domestic and international). Once the pandemic passes, this product offer could be a fast track to the industry's recovery since people will be attracted more than ever by tourism options focused on health and wellness.
  - Cruise tourism circuits—from 2017 to 2019, cruise visits to Bangladesh, India, Myanmar, Sri Lanka, and Thailand increased. Once COVID-19 passes, the cruise industry is expected to recover in these countries and globally. Because cruises involve shore excursions, these could include day visits to circuits in each country. Sri Lanka and India, for example, operated a cruise circuit between the two countries before the pandemic. River cruises in and between India and Bangladesh also offer prospects for maritime tourism circuits.
  - Ecotourism-based circuits were popular across BIMSTEC before COVID-19. Bhutan and Nepal are both expected to make strong recoveries in this segment because this type of circuit enables activities with few people and an escape to nature.
- Use the existing BIMSTEC $70,000 Tourism Fund for joint marketing activities, including setting up of the virtual BIMSTEC Tourism Information Center and repository and app.
- Create a BIMSTEC travel card, as described in the 2006 BIMSTEC Tourism Action Plan. This could be helpful for visa facilitation and immigration procedures. As recommended in the plan, a group of experts could be convened to consider this. The Asia-Pacific Economic Cooperation Business Travel Card offers a possible model for this and should be studied further. Such a card could be self-funding if combined with existing travel cards, such as Priority Pass and DragonPass, which provide airport club access. DragonPass has 30 million members and has strategic partnerships with Visa and China CITIC Bank in the PRC that reaches higher-spending travelers. A travel card could also be combined with another action plan item. BIMSTEC-wide parity in entrance fees for archaeological sites, and this could be extended to other types of cultural sites and attractions.
- Have a digital marketing focus that encourages the uploading of virtual audio and video tours and experiences from each BIMSTEC country that have cross-postings via social media, such as Facebook and YouTube. Experiences could include cooking, dance, and music lessons and performances, and interactive cultural experiences.
- Familiarization tours in each country for media (print, broadcast, online), social media influencers, and tour operators from BIMSTEC countries and generating markets. These tours were recommended in the 2006 Plan of Action for Tourism Development and Promotion for the BIMSTEC Region and will be important post-COVID-19.

---

[107] Location scouts sources: https://www.locationshub.com/, https://www.linkedin.com/groups/728907/; https://locationmanagers.org/; Bhutan has been used as a location of 67 productions, Nepal 778, India more than 12,000 productions, Sri Lanka 454, Bangladesh 1,150, Myanmar 120, and Thailand 2,082, according to the IMDb film database; C. Tuclea and P. Nistoreanu. 2011. How Film and Television Programs Can Promote Tourism and Increase the Competitiveness of Tourist Destinations. *Cactus Tourism Journal.* 2 (25).

- Student exchanges among BIMSTEC, as recommended in the 2006 Plan of Action for Tourism Development and Promotion for the BIMSTEC Region to encourage "cultural and artistic excellence." These exchanges could also leverage the growing pre-COVID-19 market trend of student and youth travel.
- A BIMSTEC tourism marketing template focused on circuits could be a basis for marketing and communications plans with governments and private entities. This could help reassure international operators and potential visitors that destinations in the BIMSTEC region follow health protocols and are safe.

4.8   Connection to 2006 BIMSTEC Tourism Action Plan: The recommended action areas of this strategic objective help accomplish the following 2006 recommendations: BIMSTEC Information Center, BIMSTEC Tourism Working Group, tour packages, tourism fund, BIMSTEC travel facilitation via a business travel card, familiarization tours, and student exchanges.

These action areas could contribute to achieving the following SDGs:

Strategic Objective 5: Focus on human resources development

To address issue area 6, focus on human resources development for meeting public and private sector tourism development needs as BIMSTEC tourism sectors plan their post-pandemic recovery and reopening. This strategic objective will focus on areas of cooperation at the regional level, where each country's human resource needs could benefit.

Recommended actions

5.1   The 2006 Plan of Action for Tourism Development and Promotion for the BIMSTEC Region suggested BIMSTEC countries could exchange sample tourism education curricula and arrange visits to one another's travel and tourism facilities. This was suggested as part of the BIMSTEC Information Center, and it could be included in the online repository and app.

5.2   Develop "tourism consciousness" or awareness programs to help local communities become more engaged in circuit development, since their offers can help enhance the experiential elements that are globally popular. These programs could also help raise local awareness on COVID-19 mitigation measures. Thailand's tourism plan encourages the "development of tourism consciousness among Thai citizens," an activity that serves as a possible model.

5.3   Through BIMSTEC, provide a portal that offers online training and capacity building for tourism and destination management organizations via hospitality and tourism schools and, if available, university public administration programs in each BIMSTEC country. The GMS Tourism Strategy includes training for public officials in tourism, which might help inform a regional-level approach. India and Sri Lanka cooperate on tourism related human resources development and skills training, which could provide useful pointers. A needs assessment could be done through BIMSTEC to determine optimal training programs and activities.

5.4   Through BIMSTEC, introduce country representatives to the globally accepted Hazard Analysis and Critical Control Points certification for food safety and training for the ISO 45001 Occupational Health and Safety Management System.

5.5   Consider adopting the short-term training courses by the GMS. Strategic Direction 1 of the GMS tourism strategy focuses on human resource development and, specifically, supporting capacity building for public officials to "strengthen their effectiveness as destination managers … and promote entrepreneurship and expand economic opportunities for youth, particularly girls."[108]

5.6   Consider adopting the Cultural Heritage Specialist Guides Training and Certification Programme for UNESCO World Heritage Sites that the GMS Tourism Strategy has applied. The training manual is available for free.[109]

5.7   Connection to the 2006 Plan of Action for Tourism Development and Promotion for the BIMSTEC Region: All the recommendations for developing human resources in the tourism industry could be offered via the virtual BIMSTEC Tourism Information Center and online repository and app. These recommended action areas will also help tackle the plan recommendation for member countries sharing training materials.

These action areas could contribute to achieving the following SDGs:

Strategic Objective 6: Improve governance, policy, and the investment climate

Recommended actions

6.1   Visa liberalization is urgently needed across BIMSTEC countries. This is one of two top priorities for the region, the other being aiming for uniform policies on COVID-19 protocols. To the extent possible, uniform visa access needs to be established, particularly for visitors and tour operators in cross-border circuits, such as the Buddhist circuit. The first step toward visa liberalization is to take stock of the visa-free agreements that are currently in place for BIMSTEC member countries. This should be followed by an assessment of visa needs based on market priorities. For example, Bhutan received very few visitors (less than 100 annually) from Sri Lanka and Myanmar from 2014 to 2018, so these are clearly not priority markets for Bhutan (Table 8).

Reducing barriers, particularly cross-border immigration procedures such as visas, through a visa-free scheme for selected nationalities or a single visa for all would greatly facilitate travel to and through the region, and thus boost the attractiveness of cross-border circuits. As Bangladesh and India demonstrated with the liberalization of the Revised Travel Arrangement in 2014, arrivals from Bangladesh to India increased 140%—and this bodes well for the future development of cross-border circuits between the two countries.

---

[108] GMS Tourism Working Group. *GMS Tourism Sector Strategy, 2016–2025.* p. 30.
[109] Specialist Guides Training Manual: https://docplayer.net/31932971-A-training-manual-for-heritage-guides-4-th-edition.html.

A 2014 UNWTO and WTTC report found that by enacting visa facilitation policies, international tourism arrivals increased for ASEAN member states between 6 million (low impact scenario) and 10 million (high impact) tourists, an increase in total arrivals of international tourists between 3.0% and 5.1% above a baseline forecast.[110] With visa facilitation in ASEAN, the report forecast an additional $6 billion to $7 billion in international tourism receipts. Visa facilitation includes liberalized visa policies, visas on arrival, and eVisa—all of which could be assisted through regional agreements.

6.2   Make it easier to invest in BIMSTEC tourism opportunities. The first step for potential investors is to collect information on the investment climate—regulations, incentives, and opportunities. Setting up a database of this information helps facilitate investments, particularly in multi-country circuits where more than one set of investment requirements and incentives are in place. Related to this, BIMSTEC should consider adapting the ASEAN Tourism Investment Guide to inform prospective investors of opportunities, which could form the basis of a BIMSTEC tourism investment strategy.

6.3   Flag BIMSTEC policies that benefit the travel and tourism industry. Set up a special section on the BIMSTEC tourism portal highlighting news of each member state's policy protocols and best practices for COVID-19 recovery, particularly those measures related to helping the industry restart.

6.4   Restoring aviation connectivity will be a priority when it is safe to travel again. BIMSTEC countries should consider an Open Skies aviation market to attract airlines back to the region.

6.5   Regional coordination on governance issues, particularly in the context of BIMSTEC and wider regional cooperation, is lacking. BIMSTEC should be able to act on approved cooperation activities independently without requiring a duplicate consensus-building effort.

6.6   In the 2006 BIMSTEC Tourism Action Plan, the establishment of a tourism working group was recommended to decide on program priorities and proposals for further action. The working group would consist of representatives from BIMSTEC national tourism organizations and industry stakeholders. The preparation of the terms of reference for this group was begun, but not completed. This is an important topic for discussion among the member countries for considering a possible change in the consensus requirement, which seems to be impeding progress on joint activities.

6.7   Connection to 2006 BIMSTEC Tourism Action Plan. This set of action areas helps address the plan recommendations of joint investment promotion, transport access, and the BIMSTEC Tourism Working Group.

These action areas could contribute to achieving the following SDGs:

---

110   UNWTO and WTTC. 2014. *The Impact of visa facilitation in ASEAN member states.* Madrid. pp. 10–11.

Short-, Medium-, and Longer-Term Recommendations for Action

The Secretariat has an important role to play in this process. The following are the recommended priority actions to achieve based on a phased-in approach, starting with immediate short-term actions (Table 28).

Short-Term Priorities and Recommended Actions

Table 28: Summary of Short-Term Actions, 6-12 Months

Strategic Objective	Action Areas	Activities	Implementing Agency	Expected Outcome	Limitations and Constraints
SO1 Control COVID-19	Exchange best practices and policies	Virtual BIMSTEC Information Center (2006)	BIMSTEC with national tourism organizations and health authorities	BIMSTEC COVID-19 protocols and standards	Insufficient intraregional cooperation. Possible hesitation in sharing data
	Adopt BIMSTEC-wide healthy destination designations	Consider implementing WTTC Safe Travels program	BIMSTEC with national tourism organizations and health authorities	BIMSTEC COVID-19 protocols and standards	Insufficient intraregional cooperation. Possible hesitation in sharing data
SO2 Recover lost jobs and businesses	Facilitate business recovery and reemployment	Damage assessment	Ministries of tourism, labor, and commerce with industry associations	Data on losses for more precise planning	Lack of data collection capacity
	Business recovery and employment planning	Data analysis	Ministries of tourism, labor, and commerce with industry associations	Strengthening of existing businesses and reemployment	Lack of data collection and analysis capacity
SO3 Improve tourism-related infrastructure	Improve tourism-related infrastructure	Add member plans to BIMSTEC Virtual Information Center (2006)	BIMSTEC	Knowledge sharing of plans, policies, and strategies	BIMSTEC Secretariat needs support on staff to maintain the center
	PPP for improved infrastructure	Identify opportunities such as roadside amenities	Member focal points obtain project opportunities from national investment agencies	Central repository of PPP tourism infrastructure opportunities	Coordination of focal point efforts

continued on next page

Table 28 continued

Strategic Objective	Action Areas	Activities	Implementing Agency	Expected Outcome	Limitations and Constraints
	Focus on Infrastructure for circuits	Begin process of determining needs	National tourism organizations with transport ministries	List of infrastructure needs for circuit development in each member country	Field visits will be difficult for needs assessments until COVID-19 subsides, which might not be until 2022 or 2023
SO4 Stimulate circuit development and marketing	Stimulate domestic and local travel	Inventory of opportunities and marketing campaigns in each member country	National and subnational tourism organizations and BIMSTEC	Comprehensive online directory of circuit opportunities and marketing campaigns	Circuits should be tested before posting. In the short and medium-term, this might be impossible due to COVID-19 and border restrictions
		Post inventory on BIMSTEC Virtual Information Center	BIMSTEC Secretariat	Comprehensive online directory of circuit opportunities and marketing campaigns	BIMSTEC Secretariat needs support on staff to maintain the center
SO5 Focus on human resources development	Tourism workforce training for restart periods	Conduct a needs assessment to determine best practices on training	Ministries of tourism, labor, and commerce with industry associations	Human resource gaps and needs identified, which can guide training	Persuading the unemployed to start or return to tourism when opportunities have not yet restarted
		Adopt and adapt GMS examples	Ministries of tourism, labor, and commerce with industry associations	GMS-tested programs introduced	Adoption and adaptation might require some time and require localization
SO6 Quality governance, policy, and investment	Reinforce BIMSTEC Tourism Working Group	Establish terms of reference for the group	Member country national tourism organizations	Group strategy	Achieving consensus from all BIMSTEC members
	Regional emergency response and recovery group	Concept note for establishment of the group	BIMSTEC with national tourism organizations and public safety agencies	Initial regional tourism risk management strategic plan	Achieving consensus from all BIMSTEC members

BIMSTEC = Bay of Bengal Initiative for Multi-Sectoral Technical and Economic Cooperation, GMS = Greater Mekong Subregion, PPP = public–private partnership, SO = strategic objective, WTTC = World Travel & Tourism Council.

Source: WTTC and Asian Development Bank.

The immediate priorities are:

- Business and employment recovery.
- Assisting in the recovery of the many businesses, especially small and medium-sized enterprises that directly and indirectly provide travel and tourism related services and goods.
- Reemploying those who lost their jobs due to the drop in demand.
- Tourism becoming a rapid catalyst and boost for economic growth in related sectors, such as construction, agriculture, cleaning services, and transportation.
- Providing the information that can help achieve the immediate priorities.

The short-term recommended actions for 2022 to realize these priorities are:

- **Set up the virtual BIMSTEC Tourism Information Center and online repository and app.** Information will be essential for enabling regional cooperation and realizing the strategic objectives and recommended action areas. The repository and app should provide direct access to information and resources to assist member countries and their travel and tourism industries to recover from COVID-19 and rebuild as quickly as possible. They should enable members to submit information and examples of their own practices easily and quickly.
- **Facilitate the recovery of businesses and lost jobs.** While all BIMSTEC countries are working on this and trying to produce solutions, an assessment to determine which businesses and workers have been most affected is an immediate priority for recovery planning. Aggregating solutions and exchanging best practices among members can inform each other on which solutions are proving to be the most effective.
- **Infrastructure.** Most improvements in tourism and connectivity infrastructure will take time to realize, but health and safety are immediate needs. Ensuring the health and safety of citizens, visitors, and businesses will be a vital part of tourism planning and development. Through the online repository and app, BIMSTEC could be as a portal for global and BIMSTEC best practices on tourism-related health and safety measures. The WTTC's Safe Travels program is worth considering, especially as authorities and tourism associations in India, Sri Lanka, and Thailand use this program.
- **Product and marketing.** International travel to and intraregional travel within BIMSTEC was expected to be on hold well into 2022. The most immediately available and accessible opportunity has been domestic and local travel. It is recommended that this should be further encouraged.
  - BIMSTEC facilitates the exchange of members' domestic marketing campaign strategies via the repository and online roundtable sessions.
  - In preparing and positioning for recovery, each member conducts an inventory of existing nature-based and cultural assets that comprise or could comprise circuits. This information could immediately be posted to the repository and app.
  - BIMSTEC facilitates the exchange of measures to foster confidence among prospective visitors and citizens.
- **Human resources.** The interim period ahead of the recovery in the tourism industry could be used for training and education.
  - Conduct a needs assessment survey in BIMSTEC countries to identify the best practices that could be exchanged and further developed.
  - Consider adopting and adapting some of the short-term training courses of the GMS for immediate application.

- **Governance, policy, and investment-related**
  - Prepare terms of reference for establishing a BIMSTEC Tourism Working Group to decide on future work priorities, as recommended in the 2006 BIMSTEC Tourism Action plan. The group should review how to expand the Secretariat's capacity to handle tourism.
  - BIMSTEC countries should consider setting up a BIMSTEC Travel and Tourism Industry Emergency Response and Recovery Group, perhaps as part of the virtual center.

Medium-Term 2024–2025

Table 29: Summary of Medium-Term Actions, 2023–2024

Strategic Objective	Action Areas	Activities	Implementing Agency	Expected Outcome	Limitations and Constraints
SO1 Control COVID-19	Exchange best practices and policies. (2006)	Continue developing the virtual BIMSTEC Information Center (2006)	BIMSTEC with national tourism organizations and health authorities	BIMSTEC COVID-19 protocols and standards	Insufficient intraregional cooperation. Possible hesitation in sharing data
	Adopt BIMSTEC-wide healthy destination designations	Continue implementing WTTC Safe Travels program	BIMSTEC with national tourism organizations and health authorities	BIMSTEC COVID-19 protocols and standards	Insufficient intraregional cooperation. Possible hesitation in sharing data
SO2 Recover lost jobs and businesses	Facilitate business recovery and reemployment (2006: Crisis Management)	Develop online information center as an online training portal. (2006)	Ministries of tourism, labor, and commerce with industry associations	Online courses offered that lead to certification and jobs	Matching students and jobs might be challenging in those countries that have not yet recovered within this period
SO3 Improve tourism-related infrastructure	Improve tourism-related infrastructure (2006: Transport access)	Identify priority actions in the Transport Connectivity Master Plan for each member	BIMSTEC and member country national tourism agencies and transport authorities	Action for tourism-related transport access improvements	Budget for transport improvements might be a constraint
		Route assessments to prioritize route pairings for a BIMSTEC-wide tourism marketing strategy	BIMSTEC and member country national tourism agencies and aviation authorities	Return of low-cost carriers and increased flights between and to BIMSTEC countries	Reassuring travelers of flights being healthy and low risk
	PPPs for improved infrastructure	Continue identifying and refining opportunities such as roadside amenities	Member focal points obtain project opportunities from national investment agencies	PPP tourism infrastructure projects initiated	Coordination of focal point efforts

continued on next page

Table 29 continued

Strategic Objective	Action Areas	Activities	Implementing Agency	Expected Outcome	Limitations and Constraints
	Focus on infrastructure for circuits (2006)	Organize virtual workshops on circuit development	National tourism organizations with transport ministries	Proposals for infrastructure improvements on circuits	Cross-border circuits might not a priority for all BIMSTEC members
SO4 Stimulate circuit development and marketing	Stimulate domestic and local travel and limited international travel	Reconstituted BIMSTEC Tourism Working Group begins working on a regional tourism strategy (2006)	National tourism organizations and BIMSTEC	First draft of regional tourism strategy with an emphasis on cross-border circuits	Depends on rates of COVID-19 vaccination and mitigation measures, and extent of border restrictions
		Develop a BIMSTEC tourism brand focused on thematic circuits	National tourism organizations and BIMSTEC	BIMSTEC regional tourism brand emphasizing circuits	Requires full member consensus
		Outreach strategy for connecting with film location scouts and production companies	National tourism organizations and BIMSTEC	Film location scout visits to BIMSTEC thematic circuits	Extensive competition for the attention of film and location scouts
		Sample tour packages of regional circuits with fam tours. (2006)	Tour operation associations	Increased demand for circuit tours	Marketing will be needed to recover within the medium-term
		Digital marketing training for BIMSTEC public and private stakeholder	National tourism organizations and BIMSTEC	Widening reach of online marketing campaigns	Capacity building regarding digital marketing and social media will be needed by some national and local tourism organizations
		Develop an app for accessing the information center	National tourism organizations and BIMSTEC	Widening reach of online marketing campaigns	No constraints expected
		Finalize program for using the Tourism Fund	National tourism organizations and BIMSTEC	Establishment of the information center	Full consensus needed for finalizing the program
SO5 Focus on human resources development	Tourism workforce training for restart periods	Use the virtual information center as an online portal for delivering training	Ministries of tourism, labor, and commerce with industry associations	Human resource gaps and needs filled	Ensuring that trainees are placed in jobs
		Continue adaptation of GMS examples	Ministries of tourism, labor, and commerce with industry associations	GMS-tested programs introduced	Adoption and adaptation might require some time and require localization
		Introduce HACCP and ISO 45001 training	Ministry of Tourism and Ministry of Health	Expansion of health and safety standards	Budget might be a constraint

continued on next page

Table 29 continued

Strategic Objective	Action Areas	Activities	Implementing Agency	Expected Outcome	Limitations and Constraints
SO6 Quality governance, policy, and investment	Reinforce BIMSTEC Tourism Working Group	Convene experts' group on visa liberalization	Member country national tourism organizations and visa authorities.	Increased visa liberalization and access resulting in increased tourism for BIMSTEC members	Achieving consensus from all BIMSTEC members
	Regional Emergency Response and Recovery Group	Formalize operation of the group	BIMSTEC with national tourism organizations and public safety agencies	Activation of the regional tourism risk management strategic plan	Achieving consensus from all BIMSTEC members
	Stimulate tourism investment	Establish a database of tourism investment opportunities	BIMSTEC with national tourism organizations and investment promotion agencies	Increased tourism investments in BIMSTEC countries	Investment post-COVID-19 has been reduced

BIMSTEC = Bay of Bengal Initiative for Multi-Sectoral Technical and Economic Cooperation, GMS = Greater Mekong Subregion, HACCP = Hazard Analysis and Critical Control Points, ISO = International Organization for Standardization, PPP= public–private partnership, SO = strategic objective.

Source: BIMSTEC and Asian Development Bank.

The medium-term priorities for the period 2024–2025 are:

- Identifying and prioritizing circuits for further infrastructure development, including the identifying sites for PPP-based roadside amenity complexes.
- Reestablishing markets through coordinated regional marketing and promotion, especially focusing on restarting aviation routes, digital marketing, and a uniform BIMSTEC tourism brand based on thematic circuits.
- Starting work on drawing up BIMSTEC tourism strategy that leverages digital technology and social media marketing.
- Leveraging regional experience for human resources development.
- Fully activating the BIMSTEC Tourism Working Group to decide on activities.

The medium-term recommended actions to realize these priorities are:

Infrastructure

- BIMSTEC countries and national tourism associations identify infrastructure needs for targeted circuits that could be transnational, especially focusing on PPP opportunities, such as roadside amenity complexes, hotels, restaurants, and cafes.
- Organize virtual workshops on circuit development with a possible view to adapting India's Swadesh Darshan Scheme's detailed project report submission toolkit. Apply the toolkit or an adaptation of it for assessing individual circuit opportunities overall and specifically.

- Because BIMSTEC is assisting in coordinating the implementation of the Transport Connectivity Master Plan and the Motor Vehicle Agreement, prioritize road projects that enhance cross-border circuits. The priority circuits would be recommended by each member.
- To help facilitate the restart of low-cost carriers and regular carrier routes, BIMSTEC should facilitate a route assessment being undertaken to prioritize route pairings and include these in a future BIMSTEC Tourism Strategy.

Product and marketing

- Under the direction of a reconstituted BIMSTEC Tourism Working Group, reestablish markets through coordinated regional marketing and promotion by developing a BIMSTEC Tourism Strategy.
- The BIMSTEC Tourism Strategy should include the following components:
  - A uniform BIMSTEC tourism brand based on the vision of the region as the world center for thematic circuits.
  - Digital marketing, especially focusing on positioning BIMSTEC circuits via social media and travel channels—for example, YouTube, Facebook, TikTok, Instagram, Steller.com.
  - An outreach strategy for BIMSTEC national tourism organizations to connect with film location scouts[111] and production companies.
  - Sample tour packages from tour operators emphasizing regional circuits with familiarization tours of these packages. Participants would include media and tour operators from BIMSTEC countries and generating markets.
  - A BIMSTEC tourism marketing template focused on circuits to assist governments and the private sector with developing and marketing circuits.
  - BIMSTEC facilitates the development of multiple product strategies, including wellness tourism, ecotourism, and cruise tourism.
  - Use the repository and app as a BIMSTEC tourist information center that includes virtual audio and video tours and experiences from each member state. Experiences could include virtual cooking lessons and online dance and music lessons and performances. These could be marketed via Viator and Airbnb Experiences.
- The BIMSTEC Tourism Working Group decides on the use of the Tourism Fund for marketing activities and develop a plan for expanding the budget.
- Convene a virtual meeting of the BIMSTEC Tourism Working Group, including private sector stakeholders, to discuss disbursing the $70,000 Tourism Fund to the Secretariat tourism function.

Human resources

- Develop the online repository and app as a portal that offers online training and capacity building.
- Consider introducing HACCP and ISO 45001 training.
- Develop "tourism consciousness" programs to help local communities become more engaged in circuit development.
- Through BIMSTEC, introduce the Cultural Heritage Specialist Guides Training and Certification Programme for UNESCO World Heritage Sites.

---

[111] Location scout source: https://www.locationshub.com/, https://www.linkedin.com/groups/728907/; https://locationmanagers.org/.

Governance, policy, and investment

- BIMSTEC convenes the BIMSTEC Tourism Working Group and an experts' group on visa liberalization to formulate a strategy for BIMSTEC visa liberalization and maximizing visa-free travel between BIMSTEC countries.
- BIMSTEC holds a special session to develop a BIMSTEC travel card. The session should include potential sponsors, such as credit card companies and travel cards (e.g., Priority Pass and DragonPass).
- Develop a plan for expanding the Secretariat's tourism program capacity.
- Establish a database of tourism investment information and opportunities on the repository and app site, which includes project opportunities, a BIMSTEC-wide tourism investment guide, information on each country's investment incentives, regulations, and requirements.

Longer-Term Priorities and Recommended Actions, 2026 and Beyond

Table 30: Summary of Longer-Term Priorities and Recommended Actions
(2026 onward)

Strategic Objective	Action Areas	Activities	Implementing Agency	Expected Outcome	Limitations and Constraints
SO1 Control COVID-19	Action against COVID-19 variants and other health crises	Continue maintaining the virtual BIMSTEC Information Center (2006)	BIMSTEC with national tourism organizations and health authorities	BIMSTEC COVID-19 protocols and standards, as well as other health and safety protocols	Insufficient intraregional cooperation. Possible hesitation in sharing data
	Formalize BIMSTEC-wide healthy destination designations	Safe Travels program designations are made permanent	BIMSTEC with national tourism organizations and health authorities	BIMSTEC COVID-19 protocols and standards as well as other health and safety protocols	Insufficient intraregional cooperation. Possible hesitation in sharing data
SO2 Recover lost jobs and businesses	Facilitate business recovery and reemployment (2006: Crisis Management)	Continue operating the online information center as an online training portal (2006)	Ministries of tourism, labor, and commerce with industry associations	Online courses offered that lead to certification and jobs	None anticipated
SO3 Improve tourism-related infrastructure	Improve tourism-related infrastructure (2006: Investment)	Develop PPP investment packages based on circuit opportunities	BIMSTEC and member country national tourism agencies and investment promotion agencies	Increased investment along circuits	Cross-border travel not fully open

continued on next page

Table 30 continued

Strategic Objective	Action Areas	Activities	Implementing Agency	Expected Outcome	Limitations and Constraints
SO4 Stimulate circuit development and marketing	Stimulate domestic and local travel, and international travel	Activation of the regional tourism strategy to increase BIMSTEC travel (2006)	National tourism organizations and BIMSTEC	Regional tourism strategy activated with an emphasis on cross-border circuits	Uncertain whether a regional tourism strategy will be operational by 2024
SO5 Focus on human resources development	Tourism workforce training for restart periods	Circuit development training, perhaps as an online circuit development academy	Ministries of tourism, labor, and commerce with industry associations.	Human resource gaps and needs filled	Ensuring that trainees are placed in jobs
SO6 Quality governance, policy, and investment	BIMSTEC is fully staffed for tourism assistance	Coordination with members on the creation of a regional aviation market	Member country national tourism organizations and aviation authorities	Increased flight access	Achieving consensus from all BIMSTEC members
	BIMSTEC circuits are structured as PPP opportunities	PPP opportunities are marketed as investment opportunities	BIMSTEC with national tourism organizations	Increased PPP for circuit development	Achieving consensus from all BIMSTEC members

BIMSTEC = Bay of Bengal Initiative for Multi-Sectoral Technical and Economic Cooperation, PPP = public–private partnership.
Source: BIMSTEC and Asian Development Bank.

The longer-term priorities for 2026 and beyond are:

- Tourism infrastructure is in place at sites along circuits with facilities that meet international health and safety standards and offer long-term local employment.
- BIMSTEC is positioned through coordinated marketing and promotion as the global center of thematic circuits that provide models of inclusive and sustainable tourism development.

The longer-term recommended actions to realize these priorities are:

Infrastructure

- Develop PPP investment packages based on circuit opportunities.

Product and marketing

- BIMSTEC coordinates the regional marketing and promotion of BIMSTEC-branded tourism circuits with governments and private sector tourism associations by implementing the Tourism Strategy.

Human resources

- BIMSTEC offers circuit development training, perhaps as an online circuit development academy.

Governance, policy, and investment

- BIMSTEC becomes a fully staffed regional organization with an expanded budget to develop and lead regional tourism activities.
- BIMSTEC coordinates with member countries to establish regional aviation market that expands connectivity throughout the region.
- BIMSTEC circuits are structured as PPP opportunities attracting substantial investor interest from investors in BIMSTEC countries and beyond. These opportunities would include possibilities for small and medium sized businesses, as well as larger businesses.

Appendix 1
Domestic Circuits in India

The following information about domestic circuits in Inida is derived from http://swadeshdarshan.gov.in/index.php?T

- Buddhist circuits
  - Bihar Convention Centre in Bodhgaya
  - Uttar Pradesh Buddhist circuit
  - Madhya Pradesh Buddhist circuit
  - Gujarat Buddhist circuit
  - Andhra Pradesh Buddhist circuit
  - Uttar Pradesh wayside amenities
- Coastal circuits
  - Odisha Coastal circuit
  - Goa Coastal circuit I
  - Goa Coastal circuit II
  - Maharashtra Sindhudurg Coastal circuit
  - West Bengal Coastal circuit
  - Tamil Nadu Coastal circuit
  - Andhra Pradesh Coastal circuit in Sri Potti Sriramulu Nellore
  - Andhra Pradesh Kakinada Hope Island Konaseema Coastal circuit
  - Andaman and Nicobar Islands Coastal circuit
  - Puducherry Coastal circuit
- Desert circuit
  - Rajasthan Development of Desert circuit
- Eco circuits
  - Telangana Ecotourism circuit in Mahaboobnagar District
  - Kerala Eco circuit: Pathanamthitta Gavi Vagamon Thekkady
  - Uttarakhand Integrated Development of Ecotourism circuit
  - Jharkhand Eco circuit
  - Madhya Pradesh Eco circuit
  - Mizoram Eco circuit II
- Heritage circuits
  - Telangana Heritage circuit
  - Rajasthan Heritage circuit
  - Assam Heritage circuit (Tezpur–Majuli–Sibsagar)
  - Uttar Pradesh Heritage circuit
  - Gujarat Heritage circuit: Vadnagar–Modhera and Patan
  - Puducherry Heritage circuit
  - Punjab Heritage circuit

- – Gujarat Gandhi circuit
- – Uttarakhand Heritage circuit
- – Madhya Pradesh Heritage circuit
- Himalayan circuits
  - – Himalayan circuit (Mantalai–Sudhmahadev–Patnitop)
  - – Himalayan circuit Gulmarg Baramulla Kupwara Leh
  - – Himalayan circuit Rajour--Shopian–Pulwama
  - – Construction of assets in lieu of those destroyed in floods
  - – Integrated development of tourism infrastructure projects under Himalayan circuit
  - – Himalayan circuit (Anantnag–Kishtwar–Pahalgam–Daksum –Ranjit Sagar)
  - – Himachal Pradesh Himalayan circuit
- Krishna circuits
  - – Haryana tourism infrastructures at places related to Mahabharata in Kurukshetra
  - – Rajasthan Govind Devji temple Jaipur, Khatu Shyam Ji, Sikar, and Nathdwara, Rajsamand
- Northeast circuits
  - – Meghalaya Northeast circuit
  - – Development of Northeast circuit II in Meghalaya
  - – Tripura Northeast circuit 2
  - – Northeast circuit Bomdila–Bhalukpong–Tawang
  - – Integrated Development of New Adventure Tourism
  - – Sikkim Northeast circuit II
  - – Manipur Tourist circuit: Imphal–Khongjom
  - – Sikkim Northeast Tourist circuit I
  - – Mizoram Northeast Eco circuit I
  - – Tripura Northeast circuit
- Ramayana circuits
  - – Uttar Pradesh Ayodhya under Ramayana circuit
  - – Uttar Pradesh Chitrakoot and Shringverpur under Ramayana circuit
- Rural circuits
  - – Rural circuit Malanad Malabar Cruise Tourism
  - – Bihar Gandhi circuit: Bhitiharwa–Chandraiah–Turkaulia
- Spiritual circuits
  - – Bihar Spiritual circuit in Bihar-Vaishali–Arrah–Masad–Patna–Rajgir–Pawapuri– Champapuri
  - – Bihar Kanwaria route under Spiritual circuit: Sultanganj to Deoghar
  - – Kerala Spiritual circuit Sree Anantha Padmanabhaswamy Aranmula Sabarimala Temples
  - – Uttar Pradesh Spiritual circuit 1
  - – Maharashtra Waki–Adasa–Dhapewada–Paradsinga
  - – Manipur Spiritual circuit
  - – Bihar Mandar Hill and Ang Pradesh
  - – Uttar Pradesh Spiritual circuit 2
  - – Rajasthan Spiritual circuit
  - – Development of Gorakhnath Temple (Gorakhpur), Devipatan Temple (Balrampur), and Vatvashni Temple (Domariyaganj)
  - – Uttar Pradesh Spiritual circuit III
  - – Pondicherry Spiritual circuit
  - – Kerala Spiritual circuit: Sabarimala Spiritual Tourism (Erumeli–Pampa–Sannidhanam)
- Sufi circuits

- None listed
- Tirthankar circuits
  - None listed
- Tribal circuits
  - Telangana Tribal circuit
  - Chhattisgarh Tribal circuit
  - Nagaland Tribal circuit (Mokokchung–Tuensang–Mon)
  - Nagaland Tribal circuit (Peren–Kohima–Wokha)
- Wildlife circuits
  - Assam Wildlife circuit
  - Madhya Pradesh Wildlife circuit

Appendix 2
Swadesh Darshan India Tourism: Detailed Project Report Submission Toolkit

PART A
1. **Executive Summary**
2. **Introduction**
   A. Tourism Overview (India's policy and direction, introduction to Swadesh Darshan Scheme)
   B. About the State (about state and its tourism policies/potential and Swadesh Darshan Scheme in the state)
   C. Objective of the Report
   D. Approach and Methodology
3. **Tourism Circuit**
   A. Circuit Description and Map (with distances between destinations and routes, nodes/modes of transportation, and their exit and entry points)
   B. Rationale (for selecting the sites, destinations, and circuit)
   C. About Tourist Destinations and Sites (discusses each destination/site and their tourism products)
      (i) Location and Regional Setting
      (ii) History
      (iii) Art
      (iv) Culture
      (v) Physiography and Climate
      (vi) Flora and Fauna
      (vii) Tourism product

PART B
4. **Existing Situational Analysis**
   A. Infrastructure Analysis (off-site infrastructure related to tourism in the state/circuit)
      (i) Approach and Accessibility (circuit and state-wise, condition of connections—road, rail, air, helipads, etc.)
      (ii) Wayside Amenities
      (iii) Automobile Refueling and Repair Stations
      (iv) Signage (from main location to airport, railheads, bus stations)
   B. Destination Assessment
      (i) Tourist Profiling
         1. Category of tourist (domestic, international, local, pilgrims, etc.)
         2. Footfall of tourist (annual, monthly, peak season, etc.)
         3. Carrying capacity of the destination/site (max. limit of tourists)
         4. Duration of stay

       (ii)   Tourist-Related Physical and Social Infrastructure—on-site (may be represented in form of a table)

          1.   Physical Infrastructure

             a.   Approach and accessibility (to various internal parts, if any)

             b.   Tourist information center

             c.   Reception center

             d.   Accommodation (both public and private sector)

             e.   Public amenities (street furniture, drinking water, public toilets, lighting, urban elements, cloak rooms, rain shelters, telecommunications, etc.)

             f.   Commercial activities (shops, commercial, space, cafes, restaurants, craft bazaars, etc.)

             g.   Waste management (solid and liquid)

             h.   Safety: CCTVs

             i.   Signage

          2.   Social Infrastructure

             a.   Safety and security (tourism police)

             b.   Medical facilities

          3.   Status of Tourism Product (if any)

             a.   Convention centers

             b.   Golf, theme parks, water and adventure parks

             c.   Ropeways

             d.   Shoreline

             e.   Waterfront development

             f.   Urban design solutions

             g.   Sound and light show

             h.   Museums

             i.   Exhibition centers, etc. (all other related tourism products shall be listed along with its condition)

          4.   Any Ongoing or Proposed Infrastructure Activity at Site Level

             a.   Name and typology

             b.   Funding agency

             c.   Status/condition

             d.   Strengths, weaknesses, opportunities, threats analysis

5.   Conclusion: Gap Identification and Projections

PART C

6.   Proposals and Benefits

    A.   Proposed Interventions (on/off site)

    B.   Regulatory Framework Appraisal (adherence to any framework affecting the project and explanation how it is tackled in the project, if any applicable of the following)

       (i)   Archaeology/Heritage Value Framework

       (ii)   Wildlife

       (iii)   Coastal Regulations

       (iv)   Green Building Codes

       (v)   Building by-laws

       (vi)   Integration with Regional Development Plan

       (vii)   Biodiversity

       (viii) Pollution Control Board Regulations, etc.

C. Marketing and Publicity Proposal
D. Project Cost: Abstract of Cost
E. Socioeconomic Benefits of Interventions (economy boost and job creation)

7. **Implementation Strategy**
   A. Project Structuring
      (i)   Central Share
      (ii)   State Share
      (iii) Public–Private Partnership Components (if any)
      (iv)   Corporate Social Responsibility components (if any)
      (v)   Other Funding (ADB, etc.)
   B. Stakeholders (listing of various organizations, agencies, departments responsible for conceptualizing and implementing projects—direct and indirect and their role)
   C. Implementing Agency/Agencies
   D. Schedule and Timing
   E. Risk Impact and Mitigation
      (i)   Market
      (ii)   Revenue
      (iii) Construction
   F. Monitoring Framework (state level)

8. **Operation and Maintenance Plan**
   A. Institutional Framework
   B. Financial Model

9. **Conclusion**

Appendix 3
Swadesh Darshan Circuit Projects in India

	Project Name	Theme	FY	Sanctioned Amount (₹')	Total Amount Released (₹)	Total UC Amount (₹)	Physical Progress (%)
Andhra Pradesh	Kakinada (Coastal)	Coastal	2014-15	67.84	67.83	67.84	100
Assam	Kaziranga (Wildlife)	Wildlife	2015-16	94.68	81.74	73.09	85
Chhattisgarh	Jashpur (Tribal)	Tribal	2015-16	99.00	79.20	71.30	82
Kerala	Gavi (Eco)	Eco	2015-16	76.55	61.24	63.93	99
Madhya Pradesh	Panna (Wildlife)	Wildlife	2015-16	92.22	78.78	74.85	99
Maharashtra	Sindhudurg (Coastal)	Coastal	2015-16	82.17	16.43	14.43	63
Manipur	Imphal (NE-I)	Northeast	2015-16	62.48	61.32	61.25	100
Mizoram	Thenzawl (Eco-I)	Eco	2015-16	92.26	87.65	90.24	100
Nagaland	Peren (Tribal-I)	Tribal	2015-16	97.36	77.89	73.77	99
Puducherry	Dubrayapet (Coastal)	Coastal	2015-16	85.28	61.82	45.79	83
Rajasthan	Sambhar (Desert)	Desert	2015-16	63.96	51.17	55.89	100
Sikkim	Rangpo (NE I)	Northeast	2015-16	98.05	92.77	92.50	100
Telangana	Somasila (Eco)	Eco	2015-16	91.62	87.04	77.45	93
Tripura	Agartala (NE-I)	Northeast	2015-16	99.59	79.67	51.36	65
Uttarakhand	Tehri (Eco)	Eco	2015-16	69.17	65.71	64.30	97
West Bengal	Udaipur (Coastal)	Coastal	2015-16	85.39	68.32	62.67	88
Andaman and Nicobar	Andaman (Coastal)	Coastal	2016-17	26.91	11.78	6.76	48
Assam	Tezpur (Heritage)	Heritage	2016-17	90.98	69.64	42.15	37
Bihar	Bodhgaya (Buddhist)	Buddhist	2016-17	98.73	48.69	22.11	36
Bihar	Sultanganj (Spiritual)	Spiritual	2016-17	52.35	39.76	37.01	86
Bihar	Vaishali (Tirthankar)	Tirthankar	2016-17	52.39	26.19	21.50	72
Goa	Baga (Coastal-I)	Coastal	2016-17	97.65	84.36	76.47	85
Gujarat	Vadnagar (Heritage)	Heritage	2016-17	91.42	85.07	80.00	96
Gujarat	Rajkot (Heritage)	Heritage	2016-17	71.77	62.63	44.82	68
Haryana	Kurukshetra (Krishna)	Krishna	2016-17	97.35	77.88	61.89	85
Himachal	Kiarighat (Himalayan)	Himalayan	2016-17	86.85	39.88	24.12	39

continued on next page

Table continued

	Project Name	Theme	FY	Sanctioned Amount (₹)	Total Amount Released (₹)	Total UC Amount (₹)	Physical Progress (%)
	SKICC–Dal Lake (Himalayan)	Himalayan	2016-17	90.96	74.70	51.56	81
	Ranjit Sagar (Himalayan)	Himalayan	2016-17	87.44	63.71	44.98	75
	Bhagwati Nagar (Himalayan)	Himalayan	2016-17	82.97	60.47	41.78	68
	Mantalai (Himalayan)	Himalayan	2016-17	97.82	75.11	45.76	57
	Rajouri-Shopian (Himalayan)	Himalayan	2016-17	96.38	48.19	41.53	61
	Baramulla–Ladakh (Himalayan)	Himalayan	2016-17	96.93	48.46	26.60	34
Kerala	Padmanabha (Spiritual)	Spiritual	2016-17	92.22	73.77	58.76	70
Kerala	Sabarimala (Spiritual)	Spiritual	2016-17	99.99	20.00	19.70	10
Madhya Pradesh	Gwalior (Heritage)	Heritage	2016-17	89.82	85.31	74.39	88
Madhya Pradesh	Sanchi (Buddhist)	Buddhist	2016-17	74.94	62.33	55.67	88
Manipur	Govindajee (Spiritual)	Spiritual	2016-17	53.80	43.04	36.88	83
Meghalaya	Umium (NE-I)	Northeast	2016-17	99.13	92.92	78.05	90
Mizoram	Durtlang (Eco-II)	Eco	2016-17	99.07	49.53	49.53	36
Nagaland	Mokokchung (Tribal-II)	Tribal	2016-17	99.67	78.09	60.77	86
Odisha	Tampara (Coastal)	Coastal	2016-17	70.82	52.96	38.17	69
Rajasthan	Nathdwara (Krishna)	Krishna	2016-17	91.45	45.72	64.80	75
Rajasthan	Churu (Spiritual)	Spiritual	2016-17	93.90	68.24	43.91	71
Sikkim	Linking Singtam (NE II)	Northeast	2016-17	95.32	76.26	49.88	70
Tamil Nadu	Chennai (Coastal)	Coastal	2016-17	72.26	59.66	59.09	93
Telangana	Mulugu (Tribal)	Tribal	2016-17	79.87	75.88	68.89	85
Uttar Pradesh	Mahoba (Spiritual)	Spiritual	2016-17	67.51	50.33	50.33	85
Uttar Pradesh	Chitrakoot (Ramayana)	Ramayana	2016-17	69.45	64.10	53.68	90
Uttar Pradesh	Unnao (Spiritual)	Spiritual	2016-17	68.39	54.71	52.62	80
Uttar Pradesh	Kalinjar Fort (Heritage)	Heritage	2016-17	33.18	26.54	19.27	57
Uttar Pradesh	Srawasti (Buddhist)	Buddhist	2016-17	99.97	72.55	52.26	64
Uttarakhand	Katarmal (Heritage)	Heritage	2016-17	76.32	67.63	63.47	94
Andhra Pradesh	Amravati (Buddhist)	Buddhist	2017-18	52.34	26.17	15.01	45
Bihar	Mandar (Spiritual)	Spiritual	2017-18	47.52	38.02	24.29	52
Bihar	Bhitiharwa (Rural)	Rural	2017-18	44.65	22.32	15.70	34
Goa	Colva (Costal-II)	Coastal	2017-18	99.35	64.68	43.26	47
Gujarat	Junagadh (Buddhist)	Buddhist	2017-18	28.67	17.40	12.83	66

continued on next page

Table continued

	Project Name	Theme	FY	Sanctioned Amount (₹')	Total Amount Released (₹)	Total UC Amount (₹)	Physical Progress (%)
Madhya Pradesh	Gandhisagar (Eco)	Eco	2017-18	94.61	79.70	55.73	74
Puducherry	Yanam (Spiritual)	Spiritual	2017-18	40.68	30.94	21.33	64
Puducherry	Nehru street (Heritage)	Heritage	2017-18	66.35	33.17	21.09	53
Rajasthan	Jaipur (Heritage)	Heritage	2017-18	90.92	49.80	39.14	85
Telangana	Qutub Shahi (Heritage)	Heritage	2017-18	96.89	70.61	45.30	59
Uttar Pradesh	Ayodhya (Ramayana)	Ramayana	2017-18	127.21	106.65	76.51	79
Jharkhand	Netarhat (Eco)	Eco	2018-19	52.72	15.07	0.00	7
Kerala	Malabar (Rural)	Rural	2018-19	80.37	23.77	0.00	0
Kerala	Narayan Guru (Spiritual)	Spiritual	2018-19	69.47	0.00	0.00	0
Maharashtra	Dhapewada (Spiritual)	Spiritual	2018-19	54.01	12.00	0.00	21
Meghalaya	Khasi Hills (NE-II)	Northeast	2018-19	84.97	25.49	0.00	0
Punjab	Amritsar (Heritage)	Heritage	2018-19	91.55	23.83	0.00	0
Tripura	Unakoti (NE-II)	Northeast	2018-19	65.00	0.00	0.00	0
Uttar Pradesh	Jewar (Spiritual)	Spiritual	2018-19	12.03	3.61	0.00	0
Uttar Pradesh	Gorakhpur (Spiritual)	Spiritual	2018-19	15.76	8.90	4.17	0
Sub-scheme	Wayside (UP–Bihar)	Wayside	2018-19	17.93	10.76	5.39	52
Total				**5,842.79**	**4,089.38**	**3,291.75**	

Source: Ministry of Tourism, India

Appendix 4
Global Sustainable Tourism Council and Destination Criteria

The Global Sustainable Tourism Council (GSTC) was created in 2010 to foster increased understanding of sustainable tourism practices and the adoption of universal sustainable tourism principles. The principles are represented through its Destination Criteria, the second version of which was released in December 2019. These criteria serve as baseline standards for tourism destination management and a framework for national and regional sustainability standards. Because of this, they are an important base for regional entities, such as the Bay of Bengal Initiative for Multi-Sectoral Technical and Economic Cooperation (BIMSTEC).

The GSTC Destination Criteria are organized into four main themes: sustainable management, socioeconomic impacts, cultural impacts, and environmental impacts. Each theme provides performance indicators and shows how they correspond to each of the Social Development Goals (SDGs). The performance indicators for each theme can be helpful guideposts for the tourism industry's recovery from the COVID-19 pandemic and its expansion in BIMSTEC countries, especially for establishing or expanding thematic circuits.

The subsections within the sustainable management theme include management structure and framework, stakeholder engagement, and managing pressure and change. The following SDGs 9, 11, 12, 13, 16, and 17 correspond to this subsection. The subsections within the socioeconomic sustainability theme include delivering local economic benefits and social well-being and impacts. The following SDGs correspond to this subsection: 1, 2, 3, 4, 5, 8, 9, 10, 11, 12, and 16. The subsections within the cultural sustainability theme include protecting cultural heritage and visiting cultural sites. The following SDGs 4, 11, 12, and 16 correspond to this subsection. The subsections within the environmental sustainability theme include conservation of natural heritage, resource management, and management of waste and emissions. The following SDGs 3, 6, 7, 9, 12, 11, 13, 14, and 15 correspond to this subsection. The outline is as follows:

- A. Sustainable Management
  - a. Management structure and framework
    - (i) Destination management responsibility
      1. SDG 16 and 17
    - (ii) Destination management strategy and action plan
      1. SDG 17
    - (iii) Monitoring and reporting
      1. SDG 12
  - b. Stakeholder engagement
    - (i) Enterprise engagement and sustainability standards
      1. SDG 12 and 17
    - (ii) Resident engagement and feedback
      1. SDG 11 and 17

        (iii) Visitor engagement and feedback
             1.   SDG 11 and 12
        (iv) Promotion and information
             1.   SDG 11 and 12

  c. Managing pressure and change
        (i)   Managing visitor volumes and activities
             1.   SDG 11 and 12
        (ii) Planning regulations and development control
             1.   SDG 9 and 11
        (iii) Climate change adaptation
             1.   SDG 13
        (iv) Risk and crisis management
             1.   SDG 11 and 16

B. Socioeconomic Sustainability
  a. Delivering local economic benefits
        (i)   Measuring the economic contribution of tourism
             1.   SDG 1, 8, and 9
        (ii) Decent work and career opportunities
             1.   SDG 4, 5, 8, and 10
        (iii) Supporting local entrepreneurs and fair trade
             1.   SDG 2, 8, and 12

  b. Social well-being and impacts
        (i)   Support for community
        (ii) Preventing exploitation and discrimination
        (iii) Property and user rights
        (iv) Safety and security
        (v)   Access for all

C. Cultural Sensitivity
  a. Protecting cultural heritage
        (i)   Protection of cultural assets
        (ii) Cultural artefacts
        (iii) Intangible heritage
        (iv) Traditional access
        (v)   Intellectual property

  b. Visiting cultural sites
        (i)   Visitor management at cultural sites
        (ii) Site interpretation

  c. Conservation of natural heritage
        (i)   Protection of sensitive environments
        (ii) Visitor management at natural sites
        (iii) Wildlife interaction
        (iv) Species exploitation and animal welfare

  d. Resource management
        (i)   Energy conservation
        (ii) Water stewardship
        (iii) Water quality

e.   Management of waste and emissions
   (i)   Wastewater
   (ii)   Solid waste
   (iii) Greenhouse gas emissions and climate change mitigation
   (iv)   Low-impact transportation
   (v)   Light and noise pollution

Appendix 5
Myanmar Health and Safety Guidelines for Tourism

2	Reopening (Relaxing of Lockdown and Quarantine) (June, July, August 2020)	(2.1)	Health and Safety of Travelers and Staff	• Set Standard Operating Procedures for transportation, accommodation, restaurants, tourism training schools and travel businesses	MOHT, MTF+11 Assoc
				• Issue safety certificates to the businesses which are qualified to operate as per safety and health standards by MOHS and inform them to the travelers	MOHS, MOHT
				• Set the Health and Safety Guidelines in Tourism industry by conducting a Workshop with the representatives from MOHT, MOHS, Tourism Executive Committee Members, and Stakeholders	MOHT
				• Organize health and safety training for all staff in the tourism industry	MOHT
				• Analyze policies on health insurance for all travelers coming to Myanmar	MOHS, MOHT
				• Issue health certificates for travelers and staff	MOHS
				• Dashboard in realtime the situation of the virus cases in Myanmar	MOHS
				• Establish Emergency Response Team in the workplace	MOHS, MOHT, RTCs, MTF
		(2.2)	Conducting Paid Training Programs	• Emply tourism professionals as trainers with budget honorarium, offering daily allowance to grass root tourism personnel (slow vehicles drivers, hawkers, etc.) for attending short programs on enhancing service, hygiene and hospitality, keeping the workforce on basic pay to join skill upgrade training	MOHT, RTCs, DMO, MTGA
				• Employ hospitality professionals as trainers with budget honorarium, offering daily allowance to existing and unemployed staff from hotels for attending refresher courses of each occupation on enhancing services and skills	MOHT, RTCs, DMO, MHA MHPA
				• Provide trainings and scholarship programs for management level staff	MOHT, DPs
				• Organize refreshment course for experience tour guides	MOHT, MTGA

continued on next page

Table continued

	(2.3) Marketing for New Normal Situation	• Encourage domestic tourism and provide holiday programs (focus on domestic market and nearby market)	MOHT, UMTA, DPTOA
		• Promote the COVID-19 free destinations	MOHT, MTF, MTM
		• Build traveler's trust by announcing the current situation and practicing of health and safety guidelines	MOHT
		• Promote tour itinerary and innovative tourism products by offering special rates	MTM, UMTA
		• Create stimulus package for domestic and international travelers	
		• Run the Visit Myanmar Now Digital Marketing Campaign	MOHT, MTF+ 11 Assoc
		• Formulate inbound tourism marketing plan	MOHT, MTF+ 11 Assoc
		• Formulate plans for travelers to keep the regulations at tourist sites and destinations	MTM, UMTA
		• Facilitate memorable activities for repeated inbound tourists to Myanmar	RTCs
			MOHT, MTGA
	(2.4) Promote ecommerce platform and digital payment	• Develop Ministry's ecommerce websites where they can put their products and itineraries	MOHT, MTF
		• Encourage ecommerce sales only accept electronic payment	MOHT, CBM, MTF, MTB
		• Focus on key generating markets by Digital Marketing	MOHT, MTF, MTM
		• Enforce skill upgrading trainings to operate ecommerce and digital payment in hotels and tourism sector	MOHT, MTF, MTM, MHPA

MOHT = Ministry of Hotels & Tourism, MTF = Myanmar Tourism Federation, MTGA = Myanmar Tour Guide Association, MTM = Myanmar Tourism Marketing, MHPA = Myanmar Hospitality Professionals Association, UMTA = Union of Myanmar Travel Association, MTB = Myanmar Tourism Board, CBM = Central Bank of Myanmar, RTC = Regional Tourism Committee, DP = development partner, MOHS = Ministry of Health and Sport.

Source: Government of Myanmar.

Appendix 6
Sample Destination Assessment Forms

The following forms are from the 2005 publication, *Linking Communities, Tourism and Conservation—A Tourism Assessment Process*, developed by Conservation International and the George Washington University for field practitioners to perform a rapid assessment and analysis of tourism potential in a destination.

Natural Attractions

WORKSHEET 01

Instructions: Describe what is unique about the natural attractions in the area. Try to be specific and avoid general attraction descriptions such as "the tropical forest."

Note: The last column asks you to choose potential market draw. This means the type of visitation an attraction may draw and may be modified based on major markets in the destination.

NATURAL ATTRACTIONS Name, Description, and Current Draw	DESCRIBE LOCATION (distance from central point or use GPS)	EASE OF ACCESS (from main entry way)	DESCRIBE POTENTIAL USES	ENVIRONMENTAL FRAGILITY (i.e. endangered species nesting area, rare plant, water source, over-crowding, waste mgmt)	SOCIO-CULTURAL CONCERNS (i.e. traditional uses and beliefs, taboos, potential disruption, land-tenure issues)	CHOOSE POTENTIAL MARKET DRAW
# __		Easy (up to 1 hr walk) Moderate (hills, 1-2hrs) Difficult (steep climbs, 2+hrs)				Day Trip Weekender Long-Stay
# __		Easy Moderate Difficult				Day Trip Weekender Long-Stay
# __		Easy Moderate Difficult				Day Trip Weekender Long-Stay
# __		Easy Moderate Difficult				Day Trip Weekender Long-Stay

Cultural Attractions

Instructions: When describing cultural attractions, express what is unique about the attractions and try to avoid general attraction descriptions such as "story telling." Note: The last column asks you to choose potential "markets." This means the type of visitation an attraction may draw and may be modified based on major markets in the destination.

CULTURAL ATTRACTIONS Name, Description, and Current Draw	LOCATION AND TIMING (when and how often)	DESCRIBE POTENTIAL ACTIVITIES	ENVIRONMENTAL FRAGILITY (i.e. endangered species nesting area, rare plant, water source, over-crowding, waste mgmt)	SOCIO-CULTURAL CONCERNS (i.e. traditional uses and beliefs, taboos, potential disruption, land-tenure issues)	CHOOSE POTENTIAL MARKET DRAW
# ___					Day Trip Weekender Long-Stay
# ___					Day Trip Weekender Long-Stay
# ___					Day Trip Weekender Long-Stay
# ___					Day Trip Weekender Long-Stay

Historic and Heritage Attractions

Instructions: When describing historical and heritage attractions, express what is unique about them and try to avoid general attraction descriptions such as "monument." Note: The last column asks you to choose potential "markets." This means the type of visitation an attraction may draw and may be modified based on major markets in the destination.

HERITAGE AND HISTORIC ATTRACTIONS Name, Description, and Current Draw (historical context)	DESCRIBE LOCATION AND ACCESS (distance from central point or use GPS)	DESCRIBE CONDITION OR RENOVATION WORK REQUIRED TO ACCOMMODATE VISITORS.	ENVIRONMENTAL FRAGILITY (i.e. endangered species nesting area, rare plant, water source, over-crowding, waste mgmt)	SOCIO-CULTURAL CONCERNS (i.e. traditional uses and beliefs, taboos, potential disruption, land-tenure issues)	CHOOSE POTENTIAL MARKET DRAW
#____					Day Trip Weekender Long-Stay
#____					Day Trip Weekender Long-Stay
#____					Day Trip Weekender Long-Stay
#____					Day Trip Weekender Long-Stay

Source: Conservation International and the George Washington University. 2005. *Linking Communities, Tourism and Conservation—A Tourism Assessment Process*. Washington, DC. 2005.

Appendix 7
BIMSTEC Airport Projects

Several airport upgrades have been made in BIMSTEC countries, including the following, which are reported in ADB's 2018 report *Updating and Enhancement of the BIMSTEC Transport Infrastructure and Logistics Study*:

- Bangladesh: Upgrading of runway at Dhaka airport 2015–2018
- Bangladesh: Improving parking aprons at Dhaka airport 2014
- Bhutan: Expansion and development of facilities at Paro airport 2014–2017
- India: Further development of Delhi airport 2014–2018
- Myanmar: Upgrading Yangon airport 2014–2016
- Nepal: Major development of Kathmandu airport 2014–2017
- Nepal: Gautam Buddha Regional International Airport in Bhairahawa 2022
- Nepal: Pokhara Regional International Airport (under construction as of July 2022)
- Sri Lanka: Phase II development of Bandaranaike International Airport, Colombo, 2015–2017
- Thailand: Major development of Suvarnabhumi Airport in Bangkok 2014–2020

Some of these projects would likely benefit from the development or enhancement of tourism circuits throughout BIMSTEC. Further research is needed to check the progress of this list.

Appendix 8
Road and Rail Projects

A. Road Projects

- Bangladesh: 4-lane Daudkandi–Chittagong highway 2014–2015
- Bangladesh: Construction of second Katchpur, Megna, Gomti bridges 2014–2018
- Bangladesh: 4-lane Benapole to Jessore 2016–2020
- Bangladesh: 4-lane Jessore to Magura to Daulatdia 2016–2020
- Bangladesh: Construction of the Padma bridge 2015–2020
- Bangladesh: 4-lane Paturia to Nabinagar 2016–2020
- Bhutan: Chhukha–Damchu bypass on Thimphu–Phuentsholing highway 2015–2016
- India: 4-lane Motihari–Raxaul National Highway 28A 2014–2015
- India: 4-lane NH from Dumdum to Barasat 2014–2018
- India: 4-lane NH from Barasat to junction State Road 2014–2018
- India: 4-lane elevated road to Kolkata port 2014–2016
- India: 4-lane access roads to Diamond Harbor 2014–2016,
- India: 4-lane missing highway link near Siliguri NH 31D 2014–2016
- India: 2–4-lane NH from Imphal to Moreh 2015–2018
- India: Improvements in highway links in West Bengal and Bihar 2014–2016
- India: 4-lane Kolkata–Siliguri corridor NH 34 2014–2020
- India: 4-lane Siliguri–Guwahati NH 31C 2014–2018
- India: 4-lane Guwahati–Shillong NH 40 2014–2015,
- Myanmar: New border link Mae Sot/Myawaddy 2015–2018
- Myanmar: Myawaddy–Kawkareik road 2014–2017
- Myanmar: Construction of Kawkareik–Eindu road 2015–2018
- Myanmar: Improvement of Thilawa–East Dagon road 2015–2017
- Myanmar: Yagyi–Kalewa road improvement 2015–2017
- Myanmar: Bridges on Kalewa–Tamu road 2014–2016
- Nepal: Connection road Birgunj 2016–2017
- Nepal: Kathmandu–Terai Fast Track Road 2016–2024
- Nepal: Nijgadh–Pathalaiya–Raxaul road upgrade 2016–2019
- Nepal: SASEC Highway Enhancement Project (Kakarbhitta-Laukahi) (To be approved in 2022)
- Sri Lanka: Port Access Expressway project 2016–2019
- Sri Lanka: Extension of Colombo–Katunayake Expressway 2015–2018
- Thailand: 4-lane of the Tak–Mae Sot highway 2014–2018
- Thailand: New border link Mae Sot, Myawaddy 2015–2018
- Thailand: Development of the Nong Kham interchange 2014–2018

B.   Rail Projects
- Bangladesh: Tongi–Bhairab Bazaar extra tracking 2014–2015
- Bangladesh: Second bridges at Bhairab Bazaar and Titas 2014–2016
- Bangladesh: 2 more lines Dhaka–Tongi and Tongi–Joydevpur 2014–2015
- Bangladesh: Double tracking Laksham–Akhaura link 2016–2019
- Bangladesh: Bridge parallel to Bangabandhu bridge 2016–2020
- India: Eastern Dedicated Freight Corridor 2014–2019
- Nepal: 5 new rail connections with India 2014–2020
- Thailand: Chachoengsao–Klong 19–Kaeng Khoi project 2014–2016

Appendix 9
Maritime Projects

- India: Elevated expressway into Chennai Port 2014–2015
- India: Additional harbor cranes at Kolkata Port 2014
- Myanmar: New port facilities at Thilawa special economic zone 2014–2020
- Sri Lanka: Extension of East Terminal Colombo 2014–2017
- Sri Lanka: Construction of West Terminal Colombo 2018–2020
- Thailand: Development of Phase III at Laem Chabang 2017–2024
- Thailand: Development of new coastal terminal at Laem Chabang 2015–2016
- Thailand: Development of new rail terminal at Laem Chabang 2015–2016

Appendix 10
BIMSTEC Country Strengths, Weaknesses, Opportunities, and Threats Tables

Table A10.1: Bangladesh Strengths, Weaknesses, Opportunities, Threats Analysis

Strengths	Weaknesses
The Seventh Five-Year Plan cited tourism as a source of "high value-added and high-income jobs" and "creating a competitive tourism industry, including ecotourism and marine cruises" as an "appropriate program." Tourism has been similarly featured in the Eighth Five-Year Plan.[a] In the 10th Islamic Conference of Tourism Ministers, Dhaka city was nominated as Organization of Islamic Cooperation City of Tourism for 2019. The Perspective Plan of Bangladesh, 2021–2041, stresses tourism as a strategic sector for the country's development.	
Infrastructure	
The country's overall development plans (Vision 2021 and Outline Perspective Plan) focus on improving infrastructure, which in turn can strengthen the tourism sector. The 2010 National Tourism Policy also recognizes the need to improve tourism infrastructure.	
Exclusive tourism zones are a priority for the government, which provide large scale eco-friendly infrastructure platforms for investment that facilitate better tourism services, in line with the targets of the Sustainable Development Goals, the National Tourism Policy 2010, the Vision 2021, the Seventh Five-Year Plan, the Delta Plan, and the Blue Economy Guidelines.[b]	
Civil Aviation in Bangladesh is in transition with growing national airline Biman Bangladesh Airlines and plans to the upgrade airport. Since 2008, Biman has acquired 16 new aircrafts from Boeing and Dash including two 787-9, four 777-300, four 787-8, two 737-800 and four Dash 8-400, which enabled the airline to operate in 8 domestic and 18 international destinations during pre-COVID-19 times. As with airlines around the world, flights were severely curtailed due to COVID-19.	

continued on next page

Table A10.1 continued

Strengths	Weaknesses
Along with increased aircraft capacity, upgrading of airports and building a new international airport made priorities in the Seventh Plan. According to the Ministry of Civil Aviation and Tourism website, improvements are planned for the following airports: Cox's Bazar Airport, Sylhet Osmani International, Hazrat Shahjalal International Airport (extension and new terminal), and Chittagong Shah Amanat International Airport. A new airport is planned for the city of Bagerhat (Khanjahan Ali Airport), a World Heritage site at the confluence of the Ganges and Brahmaputra rivers.	
Bangladesh ranks relatively strong in the 2019 WEF-TTCI in the following infrastructure indicators: • Available domestic and international seat kilometers (ranking 47th and 60th, respectively) • Road and paved road density (7th and 32nd) • Quality of railroad infrastructure and density (59th and 39th)	The following 2019 WEF-TTCI rankings underscore areas for improvements in Bangladesh's tourism infrastructure, although the government is striving for improvements in all of them: • Quality of air transport infrastructure (115th) • Number of aircraft departures (110th) • Airport density (138th) • Number of operating airlines (80th) • Quality of road infrastructure (111th) • Quality of port infrastructure (89th) • Ground transport efficiency (93rd) • Number of hotel rooms (134th) • Quality of tourism infrastructure (116th) • Presence of major car rental companies (121st) • Number of ATMs (120th)
Product and Marketing	
In a survey conducted in May 2020 with 26 Bangladeshi tour operators, 17 said they offered thematic circuits to international visitors, while 6 offer them to domestic visitors and 1 to domestic visitors only. In addition, 7 operators work with cruise lines arranging shore excursions. A visitor survey was underway as of 8 July 2020.[a]	Market data on countries of origins of visitors and reasons for visiting is lacking. The most recent reported by the United Nations World Tourism Organization are for 2014. Updated data are needed for marketing and promotion.
Bangladesh has three World Heritage Sites: the Sundarbans tiger reserves, the historic mosque city of Bagerhat, and the Ruins of the Buddhist Vihara at Paharpur. Cox's Bazar reputedly has the longest beach in the world and the Sundarbans is the world's largest mangrove forest and also part of important travel circuits for cultural and religious tourism, as well as demonstrating potential for sports tourism.	Bangladesh's ranking for indicators related to tourism products and marketing on the WEF-TCCI show improvements are needed in the following areas: • Effectiveness of marketing and branding to attract tourists (128th) • Country brand strategy (77th) • Total protected areas (102nd) • Natural tourism digital demand (113th) • Attractiveness of natural assets (126th)
With technical assistance of the South Asia Subregional Economic Cooperation Program and the ADB-funded Civil Works, Conservation Works and Awareness Programs, the South Asia Tourism Infrastructure Development Project Bangladesh 2010–2016 improved access and on-site infrastructure and facilities in the major cultural heritage sites of Kantajee Temple, Paharpur Buddhist Monastery, Mahasthangarh, and the Bagerhat Sat Gambuj mosque complex.	

continued on next page

Table A10.1 continued

Strengths	Weaknesses
A circuit for further development is a 708 kilometers (km) river cruise between India and Bangladesh that was proposed in 2019 by the India-based luxury cruise and train tour operator Exotic Heritage Group.	
Human Resources	
Several hospitality schools exist, including the National Hotel and Tourism Training Institute, which could help provide the trained personnel needed for a growing tourism sector. The Bangladesh Tourism Board is working on setting up a training institute that meets international standards to expand the skilled work force.[a] Bangladesh Tourism Board will be preparing a National Tourism Human Capital Development Strategy, 2021–2030. A consultancy firm has been awarded this assignment. The Bangladesh Tourism Board is setting up the Bangabandhu Sheikh Mujibur Rahman International Training Institute of Tourism and Hospitality. • Training programs run by board cover community-based tourism, tour operators, tourist guides, street food vendors. • United Nations Volunteer for sustainable Tourism, small and Medium Tourism Enterprise. • Service providers of tourism accommodation, Tourism transport providing opportunities to industry people for capacity building and entrepreneurship development.	According to the Sixth Five-Year Plan (FY2011–FY2015), 78% of the labor force is engaged in low-income, low-productivity work in the informal sectors. The plan states: "The quality of the labor force is weak due to low access and low quality of education; women are especially lagging behind." A human resource needs assessment has not yet been conducted.[a] The following WEF-TTCI rankings underscore areas for improvement in Bangladesh's tourism-related human resources: • Overall human resources and labor market (120th) • Qualifications of the labor force (107th) • Labor market (127th)
The National Hotel and Tourism Training Institute run by the Bangladesh Parjatan Corporation provides 2,000 well-trained graduates annually.[b]	Bangladesh has a shortage of executive-level personnel for hotel management.[b]
Governance, Policy, and Investment	
Tourism is cited as a sector for investment on the website of the Bangladesh Investment Development Authority (BIDA). The National Industrial Policy cites a tax holiday, but more information is needed. BIDA provides some technical assistance, repatriation of dividends, depreciation allowance, and tax holidays. Bangladesh's tourism policy is strong on its labor and contributions tax rate. Bangladesh ranks relatively well in the following WEF-TTCI indicators on government, policy, and investment: • Business impact of rules on foreign direct investment (52nd) • Cost to deal with construction permits (54th) • Effect of taxation on incentives to work (48th) • Effect of taxation on incentives to invest (60th) • Total tax rate (50th) • Labor and contributions tax rate (1st)	The World Trade Organization notes that there is a "serious lack of stakeholder coordination, insufficient regulatory and administrative transparency and coherence, as well as some government reluctance to relinquish greater commercial autonomy in tourism to the private sector."[c] Overall, the government seem committed to strengthening governance, which will strengthen tourism development and intraregional cooperation on tourism. The following WEF-TTCI rankings underscore the challenges in Bangladesh's tourism governance, policy, and environment, all of which the government is keen to address: • Government prioritization of travel and tourism industry (109th) • Government expenditure on travel and tourism (99th) • Openness of bilateral air service agreements (122nd) • Number of regional trade agreements in force (91st) • Property rights (89th)

continued on next page

Table A10.1 continued

Strengths	Weaknesses
Bangladesh Tourism Board plans to prepare a Tourism Master Plan for Bangladesh. An international consultancy firm has been awarded the contract to prepare this. The main target for preparing the Tourism Master Plan is to attract 10 million tourists by the year 2040 and receipts of $8 billion. To achieve these targets, 500,000 accommodations will be made available by then. The Bangladesh tourism board is going to prepare a guideline policy for the identification, preservation, and development of tourism products and services. These are: • Adventure tourism • Agro-tourism • Rural tourism • Cultural tourism • Community-based tourism • Responsible tourism • Ocean cruise tourism • River tourism • Halal tourism • Religious tourism • Sports tourism • Sustainable tourism • Wildlife tourism • Gastronomy tourism • Meetings, Incentives, Conferences and Exhibitions tourism • Blue economy tourism • Volunteerism in tourism	• Time required to deal with construction permits (129th) • Time to start a business (102nd) • Cost of starting a business (107th) • Profit tax rate (135th) • Business costs terrorism and incidence of terrorism (121st and 126th) • Use of basic sanitation and hospital beds (112th) • Ticket taxes and airport charges (108th) • Environmental sustainability (116th) • Particulate matter (140th, last) • Wastewater treatment (116th)
The country's Vision 2021 and Vision 2041 both identify tourism as a vital sector for accelerating economic growth and creating employment. A new tourism strategy is under development that accounts for the impacts of COVID-19 on tourism and increased cooperation with Bhutan and India on the Buddhist circuit.[a]	The ranking on particulate matter is particular concern and is derived from the 2018 Environmental Performance Index from Yale University, which ranked Bangladesh 179th (out of 180 countries) on 24 performance indicators across 10 categories covering environmental health and ecosystem vitality.
The Ministry of Civil Aviation and Tourism reportedly plans to introduce an e-Visa and increase the coverage of countries for visas-on-arrival.[a] In addition, visa-on-arrival is offered for the nationals of 69 countries.[a]	
The Governing Body of the Bangladesh Tourism Board is comprised of five representatives from the private sector who play a pivotal role in planning, promoting, and marketing tourism.	

continued on next page

Table A10.1 continued

Opportunities	Threats
Infrastructure	
In 2017, Bangladesh was visited for the first time by Silverseas Cruises, sparking more local interest in tapping this segment. As the Indian cruise tourism segment grows, international cruises might add Bangladesh as a destination, but this will require better port infrastructure. The government plans to develop Mongla as a cruise port.	**COVID-19** COVID-19 is the biggest threat to tourism in the BIMSTEC region. Bangladesh has been hit by cancellations of flights, hotel rooms, and tours. The government has advised against all nonessential travel. Many Bangladeshis cancelled tours to Bhutan, India, Malaysia, Singapore, Thailand, and Viet Nam. MOCAT formed the National Crisis Management Committee to prepare a recovery plan headed by the BTB chief executive officer. Standard operating procedures have been established based on UNWTO guidelines to restart inbound and domestic tourism and hospitality.
	disasters triggered by natural hazards and political violence have, in the past, taken a toll on the tourism sector. Mass protests have been frequent in Bangladesh. Bangladesh is one of the most climate vulnerable countries in the world and will become even more because of climate change, says the Ministry of Environment and Forests' Climate Change Strategy and Action Plan, 2009. Although disasters triggered by natural hazards have hindered tourism development in the past, the government and private stakeholders are doing their best to minimize the losses from future disasters.[b]
Product and Marketing	
Bangladesh has extensive potential for developing its tourism industry due to its tremendous natural and cultural resources. These include panoramic hill stations, rivers, ocean beaches, forests, a rich cultural heritage, archeological sites, historical ruins, and, above all, the hospitality of Bangladeshis.[b]	
In a May 2020 survey of 26 Bangladeshi tour operators, 15 said they wanted to offer shore excursions for cruise passengers when this segment gets back into business.	
Bangladesh and India signed a "standard operating procedure" (their term) on 25 October 2018 to operate cruise ships between the two countries. In March 2019, a Bangladesh Inland Water Transport vessel sailed from Pagla, Narayanganj and reached Kolkata in 4 days. In September, the Bangladesh newspaper *Business Standard* reported that the river vessels of the Bangladesh Inland Water Transport Corporation could not be used for river cruising, which the private sector would have to provide. A shipping ministry official said that several private companies had expressed an interest as "feasibility studies have revealed huge potential for profit from the cross-border cruise service."[d] As of mid-August 2020, Bangladesh Inland Water Transport's website did not list this voyage on its coastal vessels and inland passenger vessels routes.	

continued on next page

Table A10.1 continued

Opportunities	Threats
Other product opportunities include community-based tourism, adventure travel, sports travel, river travel, maritime tourism, and Silk Road tourism.[b]	
MOCAT is planning to link tourist destinations to multi-country circuits, i.e., Cox's Bazar, Kuakata, Kantajee Temple, Saint Martin's Island, Sonargoan, Sylhet, Sundarbans, and Ramsagor.[a] This also needs to be explored further with stakeholders.	Bangladesh lacks an internationally recognized tourism brand similar to Incredible India, making it more difficult to compete for international visitors.
Bangladesh, India, and Nepal have agreed on operating procedures for passenger vehicle movements in these countries under the Bangladesh, Bhutan, India, and Nepal Motor Vehicle Agreement signed in June 2015, and were expected to soon complete the internal approval for signing the passenger protocol. Passenger and cargo transportation by train between Bangladesh and India already exists. Tourists are moving from Bangladesh to Nepal and Bhutan via India by road. All BIMSTEC countries are well connected by air with Bangladesh. All this connectivity suggests many opportunities exist for developing multi-country tourism circuits, such as the Buddhist circuit.[b]	
According to UNWTO outbound data, India is the top source market for Bangladesh, suggesting a continuing market opportunity. More recent data are needed to take advantage of this and other market opportunities. Other top source countries include the United States, the People's Republic of China, the United Kingdom, Japan, Malaysia, the Republic of Korea, Australia, Canada, and Germany.[a]	
MOCAT was working with national tourism organizations in India and Myanmar on regional tourism promotion activities.[a]	
BTB actively develops promotional materials, such as television commercials, brochures, and other materials, for distribution worldwide. It organizes and participates in tourism fairs, events, and road shows, as well as organizes familiarization tours for foreign journalists and tour operators.[b]	
Human Resources	
The 2012 Bangladesh National Skills Survey Phase 1 estimates the tourism and hospitality industry workforce at 983,000 in 2013 and 1.1 million in 2015, with nearly 70% of workers classified as skilled and high-skilled in 2012. For the industry to grow, more trained skilled workers will be needed.	
MOCAT reports that BTB is planning to set up a training institute that follows international rules and regulations to increase the number of skilled workers in the tourism industry.[b]	

continued on next page

Table A10.1 continued

Opportunities	Threats
Governance, Policy, and Investment	
No tourism investment opportunities are listed on the BIDA website. Opportunities should be explored in and around the destinations mentioned earlier. MOCAT says the 2010 National Tourism Policy is "investment friendly."[b]	
The government has set up a tourist police force to ensure security and safety of tourists.[b]	
MOCAT has identified the following areas as potential investment opportunities: • Tourism-related infrastructure • Transportation • Amusement parks • Cable car • Cruise tourism • Sports tourism • MICE tourism	

BIMSTEC = Bay of Bengal Initiative for Multi-Sectoral Technical and Economic Cooperation; BTB = Bangladesh Tourism Board; MICE = meetings, incentives, conferences, and exhibitions; MOCAT = Ministry of Civil Aviation and Tourism; RAMSAR = The Convention on Wetlands; SLTPB = Sri Lanka Tourism Promotion Board; SWOT = strengths, weaknesses, opportunities, threats; UNWTO = UN World Trade Organization; WEF-TTCI = World Economic Forum Travel and Tourism Competitiveness Index.

[a] Reported to BIMSTEC by MOCAT 8 July 2020.
[b] Letter from Ministry of Foreign Affairs conveying comments from MOCAT on the Asian Development Bank's *Inception Report on Regional Cooperation for Tourism Development in the BIMSTEC Region*, 30 July 2020.
[c] D. Honeck and M. S. Akhtar. 2014. *Achieving Bangladesh's Tourism Potential: Linkages to Export Diversification, Employment Generation and the Green Economy*. World Trade Organization, 2014. p. 2.
[d] *Business Standard*. 2019. Bangladesh has no ships for the Dhaka-Kolkata cruise service. 4 September. https://tbsnews.net/economy/bangladesh-has-no-ships-dhaka-kolkata-cruise-service.

Table A10.2: Bhutan Strengths, Weaknesses, Opportunities, Threats Analysis

Strengths	Weaknesses
Infrastructure	
Bhutan's new tourism policy stresses the need for quality infrastructure to ensure the guiding principle of "high value, low volume" tourism. Paro international airport has flights to airports in India, Kathmandu, Dhaka, Singapore, and Bangkok.	According to the *2016 Review Report on Tourism Policy and Strategies* of the Economic Affairs Committee, only 3.6% of international tourists visited eastern Bhutan, because of poor infrastructure. The report cites the "poor quality of roads, lack of roadside amenities and long travel distances and poor quality of hotels, and the difficulties of using Gawhati as an air gateway." Tour operators and hoteliers also recognize that domestic air services are neither regular or reliable. In March 2015, Tshering Tobgay, the former Prime Minister, remarked about the "failure to develop and maintain basic tourist infrastructure like toilets." It was not clear from the review report whether another reason for the low number of visitors to eastern Bhutan is the lack of attractions in the region and whether tourism is even desired there.

continued on next page

Table A10.2 continued

Strengths	Weaknesses
Product and Marketing	
Bhutan's tourism policy, which manages the number of visitors and sets a minimum daily spending amount of $200–$250, has ensured that local culture and the environment are not disrupted and that tourism products are authentic experiences of Bhutan culture. It should be noted that service providers can charge more than this and, as the Ministry of Foreign Affairs has reported to BIMSTEC, luxury hotels do charge more.[b]	Local tour operators were reticent, according to the 2016 review report, about having their guests stay in village homestays.
Bhutan is renowned for world-class treks through Himalayan ranges and villages, as well as adventure experiences, bird watching, and spirituality and wellness experiences, especially Buddhist-related retreats and meditation.	
The Tourism Council of Bhutan (TCB) is working to diversify the country's tourism offers, including new tourism clusters, nature-based adventure, cuisine, arts, filming destination, wellness, MICE market, festival, and sports tourism. The council also intends to package the unique selling propositions of the region.	
Human Resources	
The Royal Institute for Tourism and Hospitality (RITH) opened in 2010 with assistance from the Government of Austria. Several other private institutes exist that offer catering and other professional courses, whose curricula are certified by the Ministry of Labor and Human resources and the TCB.	The TCB's 2016 *Report on Training Needs Assessment for Bhutan Hospitality and Tourism* found that nearly a third of companies had recruitment problems and labor shortages, especially for skilled chefs, cooks, bartenders, and travel service managers, and that staff turnover was high. The lack of adequate trainers and training courses, especially for short-term skills development, was widespread. The present government is planning to delink RITH from TCB with the support of the current TCB leadership. A RITH delinking task force has been formed.
The TCB under the 2019 tourism policy provides human resource development plans and skills assessments.	While the "high value, low volume" policy has been in effect for decades, the daily spending requirement of $200–$250 is outdated. The country might be able to charge far more for a high-value experience without increasing the number of visitors, and thus earn more from the same limited number. For example, Rwanda charges $1,500 per person for a 1-hour gorilla-watching excursion—and this is often fully booked. It should be noted that operators can also charge more than the minimum daily spending requirement.
Governance, Policy, and Investment	
Bhutan's tourism policy principle of "high value, low volume" tourism, which manages the number of international arrivals and requires a minimum daily spending amount, has been a model for sustainable tourism and a means for preventing over-tourism. A new comprehensive tourism policy was introduced in October 2019, which continues the "high value, low volume" policy principle. The new policy also strengthens the TCB and promises to strengthen legislation, regulatory frameworks, governance, and the workforce.	Stakeholders informed the Economic Affairs Committee in consultative sessions that a main issue was the "very weak level of organization and coordination among government agencies in regulating and promoting the sector." Stakeholders from industry associations told the committee that their industry views were not being adequately acknowledged.

continued on next page

Table A10.2 continued

Strengths	Weaknesses
The government updated the Foreign Direct Investment Policy in 2019 supporting the concept of Gross National Happiness for a green and sustainable economy, socially responsible and ecologically friendly industries, culturally and spiritually sensitive industries, promoting of Brand Bhutan, and creating a knowledge society and economic diversification—all of which are strengths for sustainable tourism development.	
The main incentives for tourism investments are a tax holiday of 10 years for newly established tourist standard hotels and 5 years for hotels upgraded to tourist standard hotels. A tax holiday of 5 years is available to farmhouse and home stays.[c]	
Exemptions from sales taxes and customs duties are on imports of buses, outdoor recreation equipment, and tourist standard hotels."[c]	
Bhutan's Government launched its 21st Century Economic Roadmap in January 2020 to guide economic development over the next 10 years. It draws on its development vision and outlines main strategies for priority economic areas.[b]	The number of international visitors to Bhutan surged from 12,500 in 2010 to 274,000 in 2018, with 39% of the market coming from East Asia and the Pacific. In 2016, when there were 200,000 fewer visitors, they raised the following issues, which might account for the drop: • Need to ensure regional visitors have access to high-quality services and experiences as provided to other international visitors. • Need to enhance safety and security of regional visitors by routing them through licensed local tour operators and tourism outlets in the southern border towns. • Ensure adherence to local laws, customs, and by using local tour guides. • Ensure road safety by using local transport operators to avoid accidents when driving in mountain roads and difficult road conditions.
An updated tourism policy includes targeted product-based incentives for promoting tourism across the country.	According to the TCB's 2016 Report on Training Needs: "Under current regulations, entry to the tourism industry and the granting of a license is comparatively easy …. Any individual who is a Bhutan national can obtain a tour operator and/or travel agent license.[d] This weakens the operator services sector.

Opportunities	Threats
Product and Marketing	
The country has an opportunity to take more advantage of the "scarcity principle" whereby limiting something, such as the number of visitors, can increase prices or value. Given other limited-number tourism experiences, such as gorilla watching treks in Africa, Bhutan has an opportunity to leverage its comparative advantages as having a limit on the number of visitors, which increases its attractiveness and value.	In 2020 and into 2021, COVID-19 was a threat to travel and tourism in Bhutan. The government acted quickly and successfully to control the pandemic's spread and was praised in international media for having only one COVID-19 death.

continued on next page

Table A10.2 continued

Opportunities	Threats
Visits to Bhutan are often combined with visits to Nepal, so more dual-country marketing is possible. Regional circuits, such as a multi-country tour of the Himalayas, are already provided by tour operators, such as HimalayaCircuit.com, which offers a tour of Bhutan, Nepal, and the western part of the People's Republic of China, and a luxury cultural tour of Nepal and Bhutan. The question of thematic trails needs to be explored further with Bhutanese officials.	
While COVID-19 will be a problem into 2022 and beyond for many generating markets for Bhutan, the domestic tourism market, wellness tourism, and transformational travel offer substantial opportunities to help the market's near-term recovery and for further developing the market.	
Human Resources	
In the new national tourism policy, the government recognizes that the Royal Institute for Tourism and Hospitality needs to be strengthened to meet the demands of a high- value tourist.	
Investment	
According to Business Bhutan, foreign direct investment policies have been revised to attract more investment, with most projects being in the hotel sector. The revised tourism policy includes product related incentives to promote more equitable growth across the country. Modest hotel investments have been made, but historically, FDI inflows have been the lowest in Asia, i.e., only $7 million in 2019.[e]	

BIMSTEC = Bay of Bengal Initiative for Multi-Sectoral Technical and Economic Cooperation; FDI = foreign direct investment; MICE = meetings, incentives, conferences, and exhibitions; RITH = Royal Institute for Tourism and Hospitality, TCB = Tourism Council of Bhutan

[a] Government of Bhutan, Economic Affairs Committee. 2016. *Review Report on Tourism Policy and Strategies.* Thimphu. p. 16.
[b] Feedback from the Ministry of Foreign Affairs, 6 July 2020.
[c] Rules on the Fiscal Incentives Act of Bhutan, 2017.
[d] Government of Bhutan, TCB. 2016. *Report on Training Needs Assessment for Bhutan Hospitality and Tourism.* Thimphu.
[e] United Nations Conference on Trade and Development. 2020. *World Investment Report.* Geneva. p. 240.

Source: Government of Bhutan and Asian Development Bank.

Table A10.3: India Strengths, Weaknesses, Opportunities, Threats Analysis

Strengths	Weaknesses
Infrastructure	
The Ministry of Tourism's website specifies "quality tourism infrastructure" as a function of the ministry. It notes that 40% of its "expenditure on Plan schemes" (such as for Swadesh Darshan) is for this purpose. The ministry provided guidelines to other government agencies in 2016 through which all tourism infrastructure could be funded. The ministry indicated that they are pursuing a scheme to help destinations and circuits through, as it says on its website, mega projects.[a] For infrastructure, it cited the following "development schemes": • Tourism product/infrastructure and destination development. • Integrated development of tourist circuits. • Assistance for large revenue-generating projects. • Capacity building for rural tourism. • Public–private partnership in infrastructure development. • Tourism market development for domestic tourism.	Roads have improved but are still considered challenging and insufficient for a growing tourism market. In the report, *Incredible India 2.0*, $1.5 trillion was estimated to be needed to improve infrastructure investments, including the number of rooms.
India did fairly well on the 2019 World Economic Forum Travel and Tourism Competitiveness Index (WEF-TTCI), ranking 33rd out of 140 countries for air transport and 28th for port infrastructure.	India ranks low in the WEF-TTCI on airport density, (airports per million population) at 135th and 109th on tourism infrastructure, in part because of a low ranking (132nd) on the number of hotel rooms.
Product and Marketing	
The government's Incredible India campaign attracted more international visitors. It also conducted the Atithi Devo Bhavah program to "sensitize people about the country's rich heritage, culture, cleanliness and warm hospitality."[a]	Tourism marketing mainly occurs at the national level via the Incredible India program and individual state-level programs. While the Incredible India program succeeded in helping to boost international arrivals, the WEF-TTCI ranks India a low 105th on country brand strategy, although better on the effectiveness of marketing (59th). Given the diversity of India's tourism offers, the country should be attracting many more visitors than it does.
The report, *Incredible India 2.0 India's $20 Billion Tourism Opportunity*, conducted by Bain and Company for the WEF shows the untapped potential for developing tourism in India's "600,00 villages with their own cultures and heritage, ecotourism and cruise tourism."[b]	
The Ministry of Tourism, under its Swadesh Darshan Scheme and the National Mission on Pilgrimage Rejuvenation and Spiritual Heritage Augmentation Drive, aimed for improved tourism infrastructure and facilities. Under the Swadesh Darshan Scheme, 15 themes have been identified for developing theme-based circuits: Buddhist circuit, Coastal circuit, Desert circuit, Eco circuit, Heritage circuit, Himalayan circuit, Krishna circuit, North-East circuit, Ramayana circuit, Rural circuit, Spiritual circuit, Sufi circuit, Tirthankar circuit, Tribal circuit, and Wildlife circuit.	
The Land of Buddha itinerary featured on the Incredible India website offers great potential for a circuit that includes Lumbini (Birthplace of Buddha) in Nepal.	

continued on next page

Table A10.3 continued

Strengths	Weaknesses
Human Resources	
The Ministry of Tourism's website indicates that the country has long had hospitality and tourism education training programs and has 41 institutes of hotel management (IHMs), comprising 21 central IHMs, 8 state IHMs, and 12 private IHMs and 5 food craft institutes following the National Council's course curriculum.	Despite a relatively large number of IHMs, Bain and Company and WEF cited human resources as an area for improvement. The WEF-TTCI ranked India at 4.5/7 for 2019 and 76th for human resources—lower than countries with less established tourism sectors, such as Benin, the Lao People's Democratic Republic, and Tajikistan.
Governance, Policy, and Investment	
India's Department of Tourism has a national tourism policy, which specifies the country's main tourism policies and development goals, objectives, and strategies. Tourism has appeared more recently in overall national economic development and planning documents. Since 2002, tourism has featured in the national development plans of the National Planning Commission. But the commission was dissolved in 2014 and does not appear to have been replaced it. Note that, as of November 2021, an updated national tourism policy was drafted.	The Department of Tourism has a national policy, but it dates back to 2002 and has not been updated. All national planning and policy efforts are challenged by a lack of coordination with the country's 29 states, each of which has its own tourism policies, plans, and strategies. These are often not coordinated with the central government.
	For the prioritization of travel and tourism, India ranked 94th on the WEF-TTCI and scored 4.3/7, partly due to the country's low ranking (130th) for travel and tourism government expenditure.

Opportunities	Threats
Infrastructure	
As the WEF-TTCI and Bain report highlight, there is still a shortage of hotel rooms in Indian tourist destinations that are expanding in the country.	As with all BIMSTEC countries, COVID-19 disrupted all inbound and domestic travel. As of December 2021, disruptions continue.
Product and Marketing	
Given the diversity of India's tourist offerings, the relatively low numbers of both international and domestic tourists (overnight visitors), as well as rapidly rising gross national income, the country's tourism product and market potential are untapped.	During normal times—and post-COVID-19—India is faced with regional competition in segments including yoga, cuisine, and cultural tourism. Political instability, which has been affecting some parts of the country, are threats to the development of the country's tourism industry, as well for establishing thematic circuits.
The Incredible India campaign, which has been used since 2001, needs refreshing. A campaign that builds on Incredible India with further emphasis of growing demand for experiential and wellness tourism could be effective both domestically and internationally. Thematic trails and circuits, especially those related to Buddhism and Hinduism, could be even more marketable when marketed jointly with Bangladesh, Bhutan and Nepal.	
Given the world's sizable Buddhist population (estimated at over half a billion), especially within Asia (over 100 million in the People's Republic of China alone), the potential attraction of a Buddhist "trail" is strong among these populations. It could also be a strong attraction for non-Buddhist cultural tourists.	

continued on next page

Table A10.3 continued

Opportunities	Threats
Human Resources	
As the tourism industry expands in India, opportunities are growing for employment in tourism and hospitality as COVID-19 recedes.	
Governance, Policy, and Investment	
Investment opportunities, particularly hotels, are being promoted along the Delhi–Mumbai Industrial Corridor.	
The country wants to attract internationally branded spa operators and wellness facilities, as well as international education and training centers. Government-owned Air India is a possible candidate for privatization.	
The WEF-TTCI ranks India 79th on business rules for FDI, 104th on the cost to deal with construction permits, 89th on how long it takes to start a new business, and 118th on the total tax rate—all disincentives to investment. India does well, however on the efficiency of its legal framework for settling disputes (32nd), challenging regulations (18th), and the effect of taxation on incentives to work (23rd) and investment (31st)—all incentives to FDI. Fostering cooperation via multi-country thematic circuits could further incentivize FDI and domestic investment.	

BIMSTEC = Bay of Bengal Initiative for Multi-Sectoral Technical and Economic Cooperation, FDI = foreign direct investment, WEF-TTCI = World Economic Forum's 2019 Travel and Tourism Competitiveness Index.

[a] K. C. Daynadra and D. S. Leelavathi. 2016. Evolution of Tourism Policy in India. *IOSR Journal of Humanities and Social Science.* 21 (11).

[b] World Economic Forum and Bain & Company. 2017. *Incredible India 2.0 India's $20 Billion Tourism Opportunity*. Geneva.

Source: WEF-TTCI

Table A10.4: Myanmar Strengths, Weaknesses, Opportunities, Threats Analysis

Strengths	Weaknesses
Infrastructure	
	The tourism master plan acknowledges that Myanmar's transport sector is underdeveloped for a country of its size, population, and potential. The plan underscores the need for improved infrastructure, and the government is aiming for improvements in roads, civil aviation, railways, cruising, as well as access to electricity, telecommunications, and health services.
	The WEF-TTCI stressed the need for improvements with the following rankings: • Quality of air transport infrastructure (136th) • Availability of international airline seat kilometers (90th) • Departures per 1,000 population (98th)

continued on next page

Table A10.4 continued

Strengths	Weaknesses
	• Quality of roads (134th) • Quality of railroad infrastructure (93rd) • Quality of port infrastructure (124th) • Quality of ground transport network (130th) • Road density (124th) • Paved road density (107th) • Hotel rooms per 100 population (124th) • Presence of major car rental companies (136th) • ATMs accepting Visa cards per million population (120th) Some of these indicators would probably now be higher given improvements since 2015, for example, the number of rooms and international airline seat kilometers.
Infrastructure has been improving. The number of hotel rooms grew from 23,500 in 2010 to 67,350 in 2018. More work is needed. See weaknesses section. The 2015 WEF-TTCI (the most recent for Myanmar) ranked the availability of domestic airline seat kilometers, number of operating airlines, and railroad density at 41st, 59th, and 60th, respectively, out of 140 countries.	
Product and Marketing	
The Tourism Master Plan, 2013–2020 highlights the country's "outstanding historic, natural, and cultural heritage" and accordingly prioritizes the following: • Cultural tourism. Festivals, heritage tours, pilgrimages, culinary tours, handicrafts, and meditation courses. • Nature-based activities. Visits to protected areas, natural sites, and beaches. • Adventure and experiential tourism. Ballooning, cycling, motorbike tours, • Kayaking and rafting, caravans, walking and trekking, volunteer tourism, and community-based tourism. • Cruise tourism and yachting. River and ocean cruises. • Meetings, incentive, conferences, and exhibitions tourism. • The 2015 WEF-TTCI cited only natural and cultural tourism digital demand (65th and 55th) as relative strengths in product and marketing.	The plan recommends that tour services be improved.
Human Resources	
The master plan recognized that workforce improvements are needed and specifies three strategic programs for making these happen. The Thirty-Year Long-Term Education Development Plan, 2001–2030 sets out directions for improving education, especially to meet the estimated needs cited in the plan of more than 563,000 trained people in tourism. The plan includes the country's first 4-year bachelor's degree in tourism and a postgraduate diploma in tourism.	The plan notes that the tourism workforce is "under significant strain to provide services that meet international standards."
Myanmar scores highly on the WEF-TTCI indicators of hiring and firing practices (51st) and women's labor force participation (19th).	

continued on next page

Table A10.4 continued

Strengths	Weaknesses
Governance, Policy, and Investment	
The policies and tourism master plan of the Ministry of Hotels and Tourism provide a comprehensive base on which to improve and sustainably develop tourism. The plan lays out a well-organized framework of guiding principles and strategic programs, which specify activities to address infrastructure, product and services, marketing and branding, human resources, and destination management.	The plan recognizes the need for better governance in tourism, stating that "Myanmar needs a wide range of measures to strengthen its institutional environment, ensure effective coordination of tourism planning and development, and achieve wider development goals."
	It lists the following measures to improve the institutional environment for tourism: • Establish a Tourism Executive Coordination Board chaired at the vice-president level. • Develop a tourism planning framework to support the board. • Strengthen tourism information systems and metrics. • Develop systems to promote visitor safety and consumer protection. • Strengthen tourism's legal and regulatory environment. Deficiencies in these areas are impeding both the development of domestic tourism and intraregional cooperation and marketing, including the possible development of thematic circuits. The following WEF-TTCI rankings underscore the wide deficiencies in Myanmar's tourism governance, policy, and environment: • Property rights (133rd) • Impact of rules FDI (122nd) • Efficiency of legal framework for settling disputes (124th) • Efficiency of legal framework for challenging regulations (126th) • Construction permits cost (125th) • Extent of market dominance (139th) • Number of days to start a business (134th) • Cost to start a business (139th) • Total tax rate (103rd) • Profit tax rate (123rd) • Other taxes rate (130th) • Government prioritization of travel and tourism industry (88th) • Government expenditure on tourism (84th) • Visa requirements (132nd)

Opportunities	Threats
Infrastructure	
Investments in new hotels and restaurants offer infrastructure-related opportunities.	Myanmar's source markets of the People's Republic of China, Japan, and the United States were all hit by COVID-19. International arrivals from outbound markets are still restricted, and all inbound travelers must undergo a 21-day quarantine.

continued on next page

Table A10.4 continued

Opportunities	Threats
Product and Marketing	
The Big 4 sites of Yangon, Mandalay, Inle lake, and Bagan attract many visitors and thus investment, particularly hotels. Outside of these destinations, there were still relatively few hotels, particularly in Kengtung, Mawlamyine, Myawaddy, Taunggyi, Sittwe, and Pyin-U-Lwin.	
River cruises offer an investment product opportunity, as do expanded coastal cruises that include Thailand and Bangladesh. A potential India–Myanmar tourist circuit could connect Bodh Gaya in Bihar state, Sanchi (Madhya Pradesh), and the upcoming capital of Amaravati (Andhra Pradesh) with Buddhist sites in Myanmar. This circuit could also include historical sites in Myanmar, which could be of interest to Indian tourists, including temples in Yangon and the grave of the last Mughal Emperor, Bahadur Shah Zafar. A Cambodia–Myanmar–India circuit is also a possiblity.[b]	
Human Resources	
In 2019, travel and tourism accounted for 4.75% of total employment in the country and more than 1 million jobs.	
Governance, Policy, and Investment	
In 2011, FDI in Myanmar's tourism sector totaled $1.14 billion, spread over 36 projects, rising to almost $3 billion in 2016, according to the Ministry of Hotels & Tourism. FDI was led by Singapore on $1.6 billion, followed by Thailand ($445 million) and Viet Nam ($440 million).	

COVID-19 = coronavirus disease, FDI = foreign direct investment, WEF-TTCI = World Economic Forum's 2019 Travel and Tourism Competitiveness Index.

[a] Government of Myanmar, Ministry of Hotels & Tourism. Traveling to Myanmar COVID-19 FAQs. https://tourism.gov.mm/travelling-to-myanmar-covid-19-faqs-moht/.

[b] T. S. Maini. 2017. Can India and Myanmar Create a Tourist Circuit? *The Diplomat*. 1 December. https://thediplomat.com/2017/12/can-india-and-myanmar-create-a-tourist-circuit/.

Source: Government of Myanmar.

Table A10.5: Nepal Strengths, Weaknesses, Opportunities, Threats Analysis

Strengths	Weaknesses
Infrastructure	
	The Nepal Tourism Vision and the WEF-TTCI identify several tourism infrastructure areas that need improving, and the current state of which is hampering the development of national and intraregional tourism.
Nepal's tourism infrastructure is weak even though the country has been welcoming tourists and pilgrims for decades. In the 2019 WEF-TTCI, only the indicators of available domestic seat kilometers (57th out of 140 countries) and airport density (20th) ranked relatively well.	The vision says Nepal's infrastructure is "insufficient," and that air connectivity and the national carrier are weak—and this is echoed in the WEF-TTCI: • Quality of air transport infrastructure (137th) • Available seat kilometers, international (79th) • Aircraft departures/1,000 population (93rd) • Number of operating airlines (74th) • Quality of road infrastructure (126th) • Road density % total territorial area (83rd) • Paved road density (75th) • Quality of port infrastructure (134th) • Ground transport efficiency (132nd) • Hotel rooms number/100 population (119th) • Quality of tourism infrastructure (80th) • Presence of major car rental companies (121st) • ATMs (115th)
Product and Marketing	
Vision 2020 underscores that Nepal's natural products are world-renowned. For decades, the Himalayas, especially Mount Everest, have been a top destination for mountaineers and trekkers. Its four UNESCO World Heritage Sites are central attractions. These are the cultural sites of the Kathmandu Valley and Lumbini, birthplace of Siddhārtha Gautama, who upon attaining enlightenment became known as "Buddha," which the Encyclopedia of Buddhism defines as "the enlightened one."; and the natural sites of Chitwan and Sagarmatha National Parks. Fifteen additional sites are on the UNESCO's tentative list. In addition, other Hindu and Buddhist sites are also popular with visitors.	Despite Nepal's rich tourism offerings, the vision notes scarce resources for large-scale publicity and consumer promotion, and that the country's tourism offerings are limited to just a few geographic areas.
The "full" traditional Buddhist circuit includes three other significant places of pilgrimage associated with the life of the Buddha. While this circuit is predominantly of interest to Buddhists, these sites also attract general tourists. Followers of Buddhism have been estimated at 500 million, or 7% of the world's population, with the majority living in Asia within easy access of Lumbini.	The WEF-TTCI highlights needed improvement areas, including on the indicators for extent of market dominance (128th) and effectiveness of marketing and branding to attract tourists (89th). All in all, however, the core product strengths of Nepal's tourism generally seem to outweigh the weaknesses.
Nepal's tourism potential, however, is constrained by numerous barriers to investment and business growth. The 2019 WEF-TTCI ranked Nepal 111th out of 140 countries on international tourism competitiveness and even lower on other rankings, such as property rights (125th). To help unlock the potential and remove constraints, The Nepal Investment Climate Reform Program commissioned a rapid assessment and scoping of tourism on the Buddhist circuit. A Buddhist trail with India offers strong potential as a two-country circuit. Bhutan, India, and Nepal are often packaged together by international tour operators.	Although tourism from third developing countries, especially for religious or spiritual purposes, to Lumbini increased by 74% from 2010 to 2019, overnight visits and positive impacts on local communities are still limited, thus limiting the significant potential to generate jobs and markets for local tourism-related goods and services in the Greater Lumbini region. This, in turn, constrains the potential for maximizing linkages for increased sales of agricultural goods to hotels and restaurants.

continued on next page

Table A10.5 continued

Strengths	Weaknesses
The WEF-TTCI ranks the country 27th on brand strategy, 32nd on World Heritage Natural Sites, 25th on protected areas, and 43rd and 46th on natural tourism and cultural tourism digital demand, respectively.	
The government has identified 100 tourist destinations for infrastructure improvements, and work on them is underway.	
The Ramayana circuit in Nepal, India, and Sri Lanka is being developed.	
More research on developing circuits is recommended.	
Human resources	
Nepal has tourism and hospitality institutes that have been providing trained personnel several years. The WEF-TTCI ranks the country 67th on its labor market indicator and 7th on women's participation in the labor force.	The vision does not identify any weaknesses in human resources, but the WEF-TTCI ranks the country as weak on the educational qualifications of the labor force (98th) and the overall labor market (67th).
Governance, Policy, Investment	
In addition to the Nepal Tourism Vision 2020, the Tourism Act specifies licensing requirements for travel and trekking agencies; hotels, restaurants, and bars; requirements and restrictions for climbing Himalayan peaks; tour guide requirements; and for setting up tourism businesses. An English language translation of the policy is needed.	Nepal is challenged by multiple issues related to policy, investment, and governance, which affect tourism development, including for intraregional circuits. The vision cites the following three issues: • Inadequate investment in tourism sector • Poor coordination among different agencies • Weak public–private partnerships
A new Tourism Act and a draft text for amending mountaineering regulations have been prepared.	
Nepal scores fairly well on the following WEF-TTCI rankings: • Time required to deal with construction permits (49th) • Effect of taxation on incentives to work (68th) • Effect of taxation on incentives to invest (61st) • Total tax rate (66th) • Rate of other taxes (25th) • Government prioritization of travel and tourism industry (41st) • Government expenditure on the industry (34th) • Visa requirements (21st)	The WEF-TTCI highlights more areas for improvement: Overall business environment (113th) Property rights (87th) Business impact of FDI rules (119th) Efficiency of legal framework in settling disputes and challenging regulations (92nd and 87th, respectively) Construction permit costs (133rd) Time it takes to start a business (89th) Cost of starting a business (109th) Labor and contributions tax rate (90th) Openness of bilateral air service agreements (132nd)

Opportunities	Threats
Nepal's rich cultural and natural assets have been attracting, as the Vision demonstrates, increased numbers of visitors from neighboring countries, as well as the "spillover effects of the Indian and Chinese markets" and the global growth in adventure travel.	COVID-19 As with all other BIMSTEC member states, COVID-19 was the biggest threat to tourism growth in the country.
Nepal's growing tourism sector also offers increasing employment opportunities. Travel and tourism generated 463,000 direct jobs and over a million total (direct and indirect) jobs, accounting for nearly 7% of all employment in 2019.	Other threats cited in the vision include: • Transitional phase of the political environment—Nepal's political environment has been in flux. • Global terrorism • Cross-border disease • Strong competing destinations • Unfavorable travel advisories • Global economic downturn • Climate change

BIMSTEC = Bay of Bengal Initiative for Multi-Sectoral Technical and Economic Cooperation, FDI = foreign direct investment.

Source: Government of Nepal and WEF-TTCI 2019.

Table A10.6: Sri Lanka Strengths, Weaknesses, Opportunities, Threats Analysis

Strengths	Weaknesses
Infrastructure	
The Sri Lanka Tourism Strategic Plan, 2017–2020 identifies the following infrastructure strengths: • Opening of significant areas that were not easily accessible during the civil war • Relatively small island with short distances between tourism sites • 61% of the national road network developed • Two international airports with planned expansions and upgrades • Development plans for domestic light aviation network • Colombo and other strategic port expansions and upgrades • Good telecommunications and internet, strong information and communication technology plans for the country	According to the plan, inconsistent policy has made developing tourism infrastructure a challenge. Infrastructure needs include highways, domestic aviation, leisure, cruise facilities, tourist jetties, and marinas. With the national policy framework, progress seems likely with more consistent policies.
Sri Lanka's infrastructure is ranked relatively strongly in the 2019 World Economic Forum Travel & Tourism Competitiveness Index (WEF-TTCI) on of the following: • Available seat kilometers, international (54th out of 140 countries) • Airport density (23rd) • Overall ground and port infrastructure (52nd) • Road density (15th) • Paved road density (50th) • Quality of railroad infrastructure (56th) • Railroad density (33rd) • Quality of port infrastructure (54th) • Quality of tourism infrastructure (35th)	The following rankings from the WEF-TTCI highlight areas of improvement in tourism infrastructure: • Overall air transport infrastructure (69th) • Quality of air transport infrastructure (82nd) • Available seat kilometers, domestic (88th) • Number of aircraft departures (83rd) • Number of operating airlines (67th) • Quality of road infrastructure (79th) • Ground transport efficiency (87th) • Overall tourist service infrastructure (92nd) • Number of hotel rooms (101st) • Presence of major car rental companies (86th) • Number of ATMs (105th)
Product and Marketing	
The country has eight UNESCO World Heritage Sites, including the ancient city of Polonnaruwa, the ancient rock fortress of Sigiriya, the old town of Galle, the sacred city of Anuradhapura, the sacred city of Kandy, the Rangiri Dambulla Cave Temple, the Central Highlands, and the Sinharaja Forest Reserve. Sri Lanka has the second-highest coverage of protected areas in Asia, with the Department of Wildlife Conservation managing nearly 90 natural areas. The Ministry of Education's Central Cultural Fund manages 21 cultural sites, including four World Heritage Sites. Sri Lanka also has several under-visited cultural and natural sites, including five RAMSAR wetlands and several Buddhist, Hindu, Christian, and Muslim religious sites that the government is working toward developing.	The plan identified the following challenges to marketing the country's tourism attractions: • Absence of a holistic approach to marketing and communications from overarching policy to strategic planning. • Absence of quality-driven, professional, digitally savvy strategic activity plans. • The Sri Lanka Tourism Promotion Board (SLTPB) focuses on limited low-return marketing activities: conventional methodologies such as trade shows, consumer shows, and above-the-line advertising. • SLTPB marketing and communications activities have not been trend conscious and dynamic in response to market requirements.
Sri Lanka cooperates with India in promoting tourism, including operating a circuit between the two, a cruise service between Kochi and Kerala, and special fares on Sri Lankan Airlines from Colombo to Kerala.	Although Sri Lanka has diverse natural and cultural attractions, its WEF-TTCI ranking on the following indicators show there is room for improvement in the following areas: • Extent of market dominance (97th) • Overall cultural resources and business travel (66th) • Oral and intangible cultural heritage (83rd) • Number of international association meetings (68th)

continued on next page

Table A10.6 continued

Strengths	Weaknesses
Sri Lanka is ranked relatively highly on the following WEF-TTCI product- and marketing-related indicators: • Country brand strategy rating (15th) • Overall natural resources (43rd) • Number of World Heritage natural sites (32nd) • Natural tourism digital demand (15th) • Attractiveness of natural assets (33rd) • Number of World Heritage cultural sites (38th) • Sports stadiums (50th) • Cultural and entertainment tourism digital demand (43rd) • Use of basic sanitation services (63rd)	
Human Resources	
	One of the tourism plan's four core strategies is to develop an actively engaged workforce in the private and public sectors. The plan also says the "SLTPB is faced with challenges in implementing and executing [tourism marketing programs] due to capability, competence and experience limitations of staff." This challenge will also be a hurdle for creating multi-country circuits.
In the Sri Lanka Tourism Strategic Plan, the government lays out plans to prioritize human resources. It advocates the greater participation in the sector of women and local communities, and in tourism training for a variety of tourism stakeholders. Sri Lanka works with Kerala Tourism in India on human resource development and skill training.	
The WEF-TTCI ranks Sri Lanka 49th on the educational qualifications of its labor force. Traditional Sri Lankan hospitality is a draw for attracting visitors.	
Governance, Policy, and Investment	
The tourism plan is a well-organized document that covers a range of issue areas in detail with specific actions and case studies. It includes a "transformational theme" of improving institutional performance, governance, and regulations." An entire chapter is devoted to improving governance and regulation, which includes: • Strong funding and empowerment of national tourism institutions in some areas (e.g., marketing). • Strong private sector entrepreneurship and institutional framework. • Large public land holding and extensive protected natural and cultural heritage areas.	The plan highlights the lack of investment and financing, and high capital costs, as challenges for further tourism development, especially for small businesses and women. It also outlines improving governance and regulation as a transformational theme through the following strategies: revitalizing institutions, improving relationships, communication, coordination, reforming core legislation and regulations, and enabling business and investment.

continued on next page

Table A10.6 continued

Strengths	Weaknesses
Tourism also features prominently in the 2019 National Policy Framework: Vistas of Prosperity and Splendor. The policy focuses on the development of tourism as an environmental and domestic culture friendly industry with extensive people's participation. Related to circuit development, the framework identifies new attractions and the development of community-based tourism and ecotourism as activities.	The plan identifies the following "challenges" related to policy and governance: • Multiplicity of government agencies with sole or shared responsibilities for important aspects of tourism. • Further fragmentation between national, provincial, and local levels of government. • Silo approach within tourism institutions, leading to inefficiency and duplication. • Inadequate planning, development, regulation, marketing, and human resource training in the public sector. • Lack of consultation, cooperation, and coordination within and between all levels of government and with the private sector. • Impediments to business and investment. • Many unregulated tourism businesses that increase risks to safety and reputation.
Sri Lanka does relatively well on the following WEF-TTCI government, policy, and investment indicators: • Time it takes to deal with construction permits (20th) • Cost to deal with construction permits (6th) • Time to start a business (57th) • Profit tax rate (11th) • Overall prioritization of travel and tourism (30th) • Government prioritization of travel and tourism industry (33rd) • Travel and tourism government expenditure (41st) • Comprehensiveness of annual travel and tourism data (46th) • Timeliness of providing monthly/quarterly travel and tourism data (24th) • Visa requirements (50th) • Hotel price index (41st) • Purchasing power parity (15th) • Sustainability of travel and tourism industry development (45th) • Particulate matter (2.5) concentration (27th) • Forest cover change (37th)	The following WEF-TTCI rankings underscore possible areas of improvement in Sri Lanka's tourism-related governance and policy: • Overall business environment (79th) • Property rights (101st) • Business impact of rules on FDI (103rd) • Efficiency of legal framework in settling disputes (77th) • Efficiency of legal framework in challenging regulations (97th) • Cost of starting a business (78th) • Effect of taxation on incentives to work (72nd) • Effect of taxation on incentives to invest (78th) • Total tax rate (122nd) • Labor and contributions tax rate (71st) • Rate of other taxes (136th) • Overall safety and security (78th) • Ticket taxes and airport charges (132nd) • Overall environmental sustainability (102nd) (policy issue) • Stringency of environmental regulations (71st) • Enforcement of environmental regulations (77th)

Opportunities	Threats
Infrastructure	
	COVID-19 As with all BIMSTEC countries, COVID-19 has been a major problem for Sri Lanka. Other threats, especially post pandemic, are competing for the same generating markets from neighboring countries or countries that offer similar segments. This is particularly true for the sun-and-sand segment.

continued on next page

Table A10.6 continued

Opportunities	Threats
According to the tourism plan, sites for developing destinations that offer infrastructure opportunities, including protected areas, wildlife safaris, whale and dolphin watching, Trincomalee natural harbor, the city and port of Hambantota, and the city of Jaffna, lack the infrastructure needed for further growth.	
Product and Marketing	
The Sri Lanka Tourism Development Authority maintains a webpage for tourism investment opportunities, which includes links for additional information on the investment process, investment opportunities, development guidelines, and development projects. The link to development projects lists seven tourism development zones and describes the Kalpitiya Integrated Tourism Resort Project.	
Kerala Tourism in India formed a tourism circuit in 2015 with Sri Lanka, which includes a cruise service between Colombo and Kochi. The cooperation also includes human resource development.	

BIMSTEC = Bay of Bengal Initiative for Multi-Sectoral Technical and Economic Cooperation, COVID-19 = coronavirus disease, FDI = foreign direct investment, RAMSAR = The Convention on Wetlands, SLTPB = Sri Lanka Tourism Promotion Board, WEF-TCCI = World Economic Forum's Travel & Tourism Competitiveness Index.

Source: Government of Sri Lanka and WEF-TTCI 2019.

Table A10.7: Thailand Strengths, Weaknesses, Opportunities, Threats Analysis

Strengths	Weaknesses
Infrastructure	
One of the five main strategies in the Second National Tourism Development Plan (NTDP), 2017–2021 of Thailand's Ministry of Tourism and Sports is the development and improvement of supporting infrastructure and amenities without inflicting negative impacts to the local communities and environment.	
Thailand's tourism infrastructure does relatively well in the following 2019 World Economic Forum Travel & Tourism Competitiveness Index (WEF-TTCI): • Overall air transport infrastructure (22nd) • Quality of air transport infrastructure (42nd) • Available seat kilometers, domestic (13th) • Available international seat kilometers (11th) • Number of aircraft departures (48th) • Number of operating airlines (9th) • Quality of road infrastructure (55th) • Road density (49th) • Paved road density (38th) • Railroad density (55th) • Overall tourist service infrastructure (14th) • Number of hotel rooms (24th) • Quality of tourism infrastructure (15th) • Presence of major car rental companies (1st) • Number of automated teller machines (13th)	The following rankings from the WEF-TTCI highlight areas for improvement in Thailand's tourism infrastructure: • Airport density (69th) • Overall ground and port infrastructure (72nd) • Quality of railroad infrastructure (73rd) • Quality of port infrastructure (66th) • Ground transport efficiency (89th)

continued on next page

Table A10.7 continued

Strengths	Weaknesses
Product and Marketing	
Two of the NTDP's other five main strategies in are (i) creating a balance between tourist target groups through targeted marketing embracing "Thainess" and (ii) creating confidence among tourists and the development of tourist products. Measures such as are outlined in the plan to develop tourist products: improving the quality of tourism offerings for all tourist segments, developing tourism offerings that are culturally and environmentally sustainable, and creating balanced development in tourism offerings in both regions.	The NTDP highlights the following measures to improve marketing: • Reinforce Thailand as a quality and safe destination • Use targeted marketing to attract and encourage visits from specific segments • Promote the uniqueness of Thailand and individual destinations • Encourage domestic tourism to create location and time balance • Strengthen marketing campaigns through stakeholder collaboration and technology.
Thailand does relatively well in the WEF-TTCI, on the following product and marketing indicators: • Effectiveness of marketing and branding to attract tourists (18th) • Overall natural resources (10th) • Number of World Heritage naturals sites (32nd) • Total known species (18th) • Natural tourism digital demand (4th) • Attractiveness of natural assets (18th) • Overall cultural resources and business travel (35th) • Number of World Heritage cultural sites (62nd) • Sports stadiums (38th) • Number of international association meetings (24th) • Cultural and entertainment tourism digital demand (22nd)	The WEF-TTCI ranks Thailand's tourism products and marketing as relatively weak in the following areas: • Country brand strategy rating (72nd) • Total protected areas (66th) • Oral and intangible cultural heritage (83rd)
Human Resources	
The last of the five main strategies in Thailand's Second NTDP 2017–2021 is the "development of tourism human capital's potential and the development of tourism consciousness among Thai citizens."	The Second NTDP outlines the following measures to improve human resources: • Enhance capabilities of those employed in the tourism industry to gain competitiveness and adhere to international standards. • Develop human resources in the tourism industry that meet market needs. • Equip local communities with capabilities to support, participate, and benefit from tourism.
Thailand does relatively well in the WEF-TTCI, on the following human resources indicators: • Overall human resources and labor market (27th) • Qualification of the labor force (26th) • Labor market (44th)	
Governance, Policy, and Investment	
The Second NTDP is a well-researched, comprehensive plan and strategy document. The Ministry of Tourism and Sports has been implementing the strategy successfully since 2017. Through this strategy, the government aims to increase private sector investment and government investment. It also outlines initiatives to "promote good management and governance to support effective tourism development."	The Second NTDP outlined the following initiatives to improve governance, policy, and investment: • Improving work integration among central entities by appointing the National Policy Committee for Tourism and Ministry of Tourism and Sports as the lead manager. • Decentralizing implementation of tourism policies by allocating and empowering local authorities across all levels by defining clear roles and responsibilities.

continued on next page

Table A10.7 continued

Strengths	Weaknesses
	• Promoting synergies and continuous best practice sharing through collaboration between public and private stakeholders across all levels. • Optimizing budget and human resources allocation to accommodate the needs of each entity across levels. • Promoting engagement with local communities in the development and management of community-based tourism. • Optimizing and facilitate tourism investment from private sector • Setting up an entity to manage and analyze tourism data for the use of relevant entities.
Thailand does relatively well in the WEF-TTCI, on the following government, policy, and investment indicators: According to the 2019 • Overall business environment (37th) • Business impact of rules on FDI (48th) • Efficiency of legal framework in settling disputes (48th) • Time required to deal with construction permits (51st) • Cost to deal with construction permits (20th) • Time to start a business (21st) • Cost of starting a business (50th) • Effect of taxation on incentives to work (50th) • Effect of taxation on incentives to invest (37th) • Total tax rate (38th) • Labor and contributions tax rate (18th)	The following WEF-TTCI rankings highlight areas for improvement in Thailand's tourism-related governance, policy, investment, and environment: • Property rights (74th) • Efficiency of legal framework in challenging regulations (71st) • Extent of market dominance (96th) • Profit tax rate (111th) • Overall safety and security (111th) • Travel and tourism government expenditure (85th) • Openness of bilateral air service agreements (84th) • Overall environmental sustainability (130th) • Stringency of environmental regulations (104th) • Enforcement of environmental regulations (92nd)
• Rate of other taxes (64th) • Use of basic sanitation services (58th) • Use of basic drinking water services (55th) • Overall prioritization of travel and tourism (27th) • Government prioritization of travel and tourism industry (7th) • Comprehensiveness of annual travel and tourism data (29th) • Timeliness of providing monthly/quarterly travel and tourism data (32nd) • Overall international openness (45th) • Visa requirements (29th) • Number of regional trade agreements in force (49th) • Overall price competitiveness (25th) • Ticket taxes and airport charges (36th) • Hotel price index (42nd) • Purchasing power parity (37th) • Fuel price levels (39th) • Sustainability of travel and tourism industry development (50th)	• Particulate matter (2.5) concentration (131st) • Threatened species (111th) • Forest cover change (82nd) • Wastewater treatment (73rd)

continued on next page

Table A10.7 continued

Opportunities	Threats
The Second NTDP specifies multiple initiatives for each strategy, which facilitates a range of tourism investment opportunities. These include: • Wellness tourism • Meetings and incentives segment • Gastronomy tourism • Shopping tourism • Sports tourism • Development of tourism "sub-clusters" in less developed parts of the country • Focus on long-terms stays from higher income visitors.	**COVID-19** As with all BIMSTEC member countries, COVID-19 stopped almost inbound and domestic tourism. Thailand has been heavily dependent on several generating markets. There seems to be no region in the world where Thailand has not successfully marketed itself—and all these markets are threatened by COVID-19 and the collapse in travel.
Human Resources	
Travel and tourism have been an important source of employment in Thailand with the industry generating 5.1 million jobs directly and a total of more than 8 million jobs.	

BIMSTEC = Bay of Bengal Initiative for Multi-Sectoral Technical and Economic Cooperation, COVID-19 = coronavirus disease, FDI = foreign direct investment, NTDP = National Tourism Development Plan, WEF-TTCI = World Economic Forum's 2019 Travel and Tourism Competitiveness Index.

Source: Government of Thailand and WEF-TTCI 2019.

Appendix 11
Sustainable Development Goals and Tourism

The following is extracted from the United Nations World Tourism Organization's 2017 publication *Tourism and the Sustainable Development Goals: Journey to 2030*.

SDG	Tourism and SDGs
SDG 1: End poverty in all its forms everywhere	Tourism provides income through job creation at local and community levels. It can be linked with national poverty reduction strategies and entrepreneurship. Low skills requirements and local recruitment can empower less favored groups, particularly youth and women.
SDG 2: End hunger, achieve food security and nutrition, promote sustainable agriculture	Tourism can spur sustainable agriculture by promoting the production and supplies to hotels, and sales of local products to tourists. Agritourism can generate additional income while enhancing the value of the tourism experience.
SDG 3: Ensure healthy lives and promote well-being for all at all ages	Tax income generated from tourism can be reinvested in health care and services, improving maternal health, reducing child mortality, and preventing diseases. Visitors' fees collected in protected areas can as well contribute to health services.
SDG 4: Ensure inclusive and equitable quality education and promote lifelong learning for all	Tourism has the potential to promote inclusiveness. A skillful workforce is crucial for tourism to prosper. The tourism sector provides opportunities for direct and indirect jobs for youth, women, and those with special needs, who should benefit through educational means.
SDG 5: Achieve gender equality and empower all women and girls	Tourism can empower women, particularly through the provision of direct jobs and income generation from micro, small, and medium-sized enterprises in tourism and hospitality-related enterprises. Tourism can be a tool for women to become fully engaged and lead in every aspect of society.
SDG 6: Ensure availability and sustainable management of water and sanitation for all	Tourism investment requirements for providing utilities can play a critical role in achieving water access and security, as well as hygiene and sanitation for all. The efficient use of water in tourism, pollution control, and technology efficiency can be important to safeguarding our most precious resource.
SDG 7: Ensure access to affordable, reliable, sustainable, and modern energy for all	As a sector, which is energy-intensive, tourism can accelerate the shift toward increased renewable energy shares in the global energy mix. By promoting investments in clean energy sources, tourism can help to reduce greenhouse gases, mitigate climate change, and contribute to accessing energy for all.

continued on next page

Table continued

SDG	Tourism and SDGs
SDG 8: Promote sustained, inclusive, and sustainable economic growth, employment, and decent work for all	Tourism, as services trade, is one of the top four export earners globally, currently providing 1 in 10 jobs worldwide. Decent work opportunities in tourism, particularly for youth and women, and policies that favor better diversification through tourism value chains can enhance tourism positive socioeconomic impacts.
SDG 9: Build resilient infrastructure, promote inclusive and sustainable industrialization and foster innovation	Tourism development relies on good public and private infrastructure. The sector can influence public policy for infrastructure upgrade and retrofit, making them more sustainable, innovative, and resource-efficient and moving toward low-carbon growth, thus attracting tourists and other sources of foreign investment.
SDG 10: Reduce inequality within and among countries	Tourism can be a powerful tool for reducing inequalities if it engages local populations and all stakeholders in its development. Tourism can contribute to urban renewal and rural development by giving people the opportunity to prosper in their place of origin. Tourism is an effective means for economic integration and diversification.
SDG 11: Make cities and human settlements inclusive, safe, resilient, and sustainable	Tourism can advance urban infrastructure and accessibility, promote regeneration, and preserve cultural and natural heritage—assets on which tourism depends. Investment in green infrastructure (more efficient transport, reduced air pollution) should result in smarter and greener cities for not only residents but also tourists.
SDG 12: Ensure sustainable consumption and production patterns	The tourism sector needs to adopt sustainable consumption and production modes, accelerating the shift toward sustainability. Tools to monitor sustainable development impacts for tourism including for energy, water, waste, biodiversity, and job creation will result in enhanced economic, social, and environmental outcomes.
SDG 13: Take urgent action to combat climate change and its impact	Tourism contributes to and is affected by climate change. Tourism stakeholders should play a leading role in the global response to climate change. By reducing its carbon footprint, in the transport and accommodation sector, tourism can benefit from low-carbon growth and help tackle one of the most pressing challenges of our time.
SDG 14: Conserve and sustainably use the oceans, seas, and marine resources for sustainable development	Coastal and maritime tourism rely on healthy marine ecosystems. Tourism development must be a part of integrated coastal zone management to help conserve and preserve fragile marine ecosystems and serve as a vehicle to promote a blue economy, contributing to the sustainable use of marine resources.
SDG 15: Protect, restore, and promote sustainable use of terrestrial ecosystems and halt biodiversity loss	Rich biodiversity and natural heritage are often the main reasons why tourists visit a destination. Tourism can play a major role if sustainably managed in fragile zones, not only in conserving and preserving biodiversity, but also in generating revenue as an alternative livelihood to local communities.

continued on next page

Table continued

SDG	Tourism and SDGs
SDG 16: Promote peaceful and inclusive societies, provide access to justice for all and build inclusive institutions	As tourism revolves around billions of encounters between people of diverse cultural backgrounds, the sector can foster multicultural and inter-faith tolerance and understanding, laying the foundation for more peaceful societies. Tourism, which benefits and engages local communities, can also consolidate peace in post-conflict societies.
SDG 17: Strengthen the means of implementation and revitalize the global partnership for sustainable development	Due to its cross-sector nature, tourism can strengthen public–private partnerships and engage multiple stakeholders—international, national, regional, and local—to work together to achieve the SDGs and other common goals. Public policy and innovative financing are at the core for achieving the 2030 Agenda.

SDG = Sustainable Development Goal.

Source: UN World Tourism Organization.

References

Adventure Travel Trade Association. 2018. 20 *Adventure Travel Trends to Watch in 2018*. https://cdn.adventuretravel
.biz/research/2018-Travel-Trends. pdf.

Asian Development Bank (ADB) 2020. *Key Indicators Database*. https://kidb.adb.org/kidb/ (accessed 31 May 2020).

ADB. 2016. *South Asia Subregional Economic Cooperation Operational Plan 2016–2025*. https://www.adb.org/sites/
default/files/institutional-document/193351/sasec-operational-plan2016-2025.pdf.

ADB. 2004. *Asian Development Bank and South Asia Subregional Economic Cooperation Tourism Working Group*.
https://www.adb.org/sites/default/files/publication/27954/tdp-final.pdf.

Association of Southeast Asian Nations (ASEAN). 2015. *ASEAN Tourism Strategic Plan 2016-2025*. Jakarta.

ASEAN. ASEAN online tourism database. https://data.aseanstats.org/dashboard/tourism (accessed
15 December 2020).

Bremner, Caroline. 2019. *Megatrends Shaping the Future of Travel*. London: Euromonitor International.

Centers for Disease Control and Prevention (CDC). *CDC COVID-19 Travel Recommendations by Destination*
https://www.cdc.gov/coronavirus/2019-ncov/travelers/map-and-travel-notices.html (accessed
17 December 2020)

CDC. *Daily Situation Summary*. https://www.cdc.gov/coronavirus/2019-ncov/cases-updates/summary.html

Constantin, Margaux, Steve Saxon, and Jackey Yu. 2020. Reimagining the $9 Trillion Tourism Economy
—What Will It Take?. *McKinsey & Company*. 5 August. https://tinyurl.com/y5pxqtde.

Daynadra, K. C. and D.S. Leelavathi. 2016. Evolution of Tourism Policy in India. *IOSR Journal of Humanities and
Social Science* 21, no. 11 (November).

Elyssa, Abby. 2018. Mia Kyricos-Hyatt Hotels Corp. *Hotel Business*. 15 December. https://www.hotelbusiness
.com/mia-kyricos-hyatt-hotels-corp/.

eTurboNews. 2020. *India Tourism Shocked: No COVID-19 Assistance*. 19 May. https://www.eturbonews.
com/572562/india-tourism-shocked-no-covid-19-assistance/.

Global Wellness Institute. 2014. *Global Wellness Tourism Economy Report 2013*. Miami.

Gupta, Aashish. 2020. Indian Tourism Industry Has to Survive – It's a Necessity. *BusinessWorld*. 6 June. https://www.businessworld.in/article/Indian-Tourism-Industry-Has-To-Survive-it-s-A-Necessity/ 06-06-2020-194489/.

———. 2018a. *What is Wellness Tourism?* https://globalwellnessinstitute.org/what-is-wellness/what-is-wellness -tourism/.

———. 2018b. *New Study Reveals Wellness Tourism Now a $639 Billion Market.* https://globalwellnessinstitute.org/ press-room/press-releases/new-study-reveals-wellness-tourism-now-a-639-billion-market/.

———. 2018c. *Global Wellness Tourism Economy.* November.

———. 2018d. *Wellness Industry Statistics & Facts.* https://globalwellnessinstitute.org/press-room/statistics-and-facts/

Government of Bangladesh, Bangladesh Tourism Board, Ministry of Civil Aviation and Tourism. 2020. *Standard Operating Procedures to restart the tourism industry during COVID-19.* August.

Government of Bangladesh, Bangladesh Tourism Board. 2020. *Recovery plan of Bangladesh Tourism Board from COVID 19.* 1 June. https://tourismboard.portal.gov.bd/site/page/7f734bbf-4185-47f3-bd9c-dec79868d54e

Government of Bhutan. 2019. *Tourism Policy of the Kingdom of Bhutan (Final Draft).* Tourism Council of Bhutan, October 2019. Thimphu.

Government of Bhutan. 2016. *Review Report on Tourism Policy and Strategies of the Economic Affairs Committee.* Thimphu.

Government of Bhutan, Ministry of Economic Affairs. 2019. *Foreign Direct Investment Policy.* Thimphu.

Government of Bhutan, Ministry of Finance. 2017. *Rules on the Fiscal Incentives Act of Bhutan.* Thimphu.

Government of Bhutan, Ministry of Health. 2020. COVID-19 Update. http://www.moh.gov.bt/national-situational -update-on-covid-19-as-of-24th-june-2020/

Government of India, Department of Tourism. 2019. *India Tourism Statistics at a Glance 2019.* Delhi.

Government of India, Department of Tourism. 2002. *National Tourism Policy.* Delhi.

Government of Myanmar, Ministry of Hotels & Tourism. 2020. *COVID-19 Tourism Relief Plan.* Yangon. https://tourism.gov.mm/wp-content/uploads/2020/06/COVID-19-Tourism-Relief-Plan.pdf

Government of Myanmar, Ministry of Hotels & Tourism. 2013. *Myanmar Tourism Master Plan 2013-2020.* Yangon.

Government of Myanmar, Ministry of Hotels & Tourism's Statistics: https://tourism.gov.mm/statistics/arrivals -2019-december/

Government of Nepal, Ministry of Tourism and Aviation. 2016. *National Tourism Strategy for 2016-2025.* Kathmandu.

Government of Sri Lanka, Ministry of Tourism Development and Christian Religious Affairs (now known as the Ministry of Tourism, 2020) *Sri Lanka Tourism Strategic Plan 2017–2020.*

Government of Sri Lanka. December 2019. *National Policy Framework 'Vistas of Prosperity and Splendor.'* Colombo.

Government of Sri Lanka. Sri Lanka Ministry of Tourism and Aviation, April 2020, *Operational Guidelines on preparedness and response COVID-19 outbreak for tourism industry*, https://srilanka.travel/covid19/pdf/Operational_Guidelines_for_the_Tourism.pdf

Government of Thailand, Ministry of Tourism and Sports. 2017. *Second National Tourism Development Plan, 2017-2021.* Bangkok.

Honeck, Dale and Md. Shoaib Akhtar. 2014. *Achieving Bangladesh's Tourism Potential: Linkages to Export Diversification, Employment Generation and the 'Green Economy.'* 26 August. Available at SSRN: https://ssrn.com/abstract=2542116 or http://dx.doi.org/10.2139/ssrn.2542116.

International Air Transport Association (IATA). COVID-19 Travel Regulations Map. https://www.iatatravelcentre.com/international-travel-document-news/1580226297.htm (accessed 5 December 2020).

Kumar, P. Krisha. 2020. Ministry of Tourism Asks States to Chart out New Circuits to Promote Domestic Travel. *ETTravelWorld.com.* 22 June. https://travel.economictimes.indiatimes.com/news/ministry/ministry-of-tourism-asks-states-to-chart-out-new-circuits-to-promote-domestic-travel/76508496.

Mekong Tourism Coordinating Office. 2017. *Greater Mekong Subregion Tourism Sector Strategy 2016-2025.* Bangkok.

Nepal24Hours Inc. 2020. Nepal Central Bank Brings Refinancing Work Procedure, 2077. 2 June. https://www.nepal24hours.com/nepal-central-bank-brings-refinancing-work-procedure-2077/

Nistoreanu, Puiu and Claudia-Elena Tuclea. 2011. How film and television programs can promote tourism and increase the competitiveness of tourist destinations. *Cactus Tourism* Journal Vol. 2, Issue 2/2011 (2011): 25.

Overseas Airline Guide. *Overseas Airline Guide Blog* (accessed 5 December 2020)

Plan of Action for Tourism Development and Promotion for the BIMSTEC Region. Second BIMSTEC Ministers' Roundtable and Workshop, Kathmandu, 2006.

Regional Government of Galicia in Spain. Build your Trip App: https://www.turismo.gal/planifica-a-tua-viaxe?langId=en_US

Skift. *Megatrends Defining Travel in 2019.* https://skift.com/2019/01/16/how-hyatt-is-taking-steps-to-incorporate-wellness-into-every-aspect-of-its-business/

Sri Lankan Airlines. Emergency news alert. https://www.srilankan.com/en_uk/coporate/emergency-news-detail/517

Statista. 2020. *Number of international tourist arrivals worldwide from 2010-2019 (in millions).* https://www.statista.com/statistics/209334/total-number-of-international-tourist-arrivals/.

Themed Entertainment Association. 2019. *TEA/AECOM 2019 Theme Index and Museum Index: The Global Attractions Attendance Report.*

Tourism Council of Bhutan. 2016. *Report on Training Needs Assessment for Bhutan Hospitality & Tourism.* Thimphu.

United Nations Conference on Trade and Development (UNCTAD). 2020. World Investment Report. Geneva.

United Nations Development Program (UNDP) and United Nations World Tourism Organization (UNWTO). 2017. *Tourism and the Sustainable Development Goals – Journey to 2030.*

UNWTO. 2020. *UNWTO/GTERC Asia Tourism Trends.* UNWTO, Madrid.

——. 2020. *UNWTO World Tourism Barometer Statistical Annex, Volume 18, Issue 7 (December).* Madrid.

——. 2020. *UNWTO World Tourism Barometer Statistical Annex, Volume 18, Issue 6 (October).* Madrid.

——. 2020. *UNWTO World Tourism Barometer Statistical Annex, Volume 18, Issue 3 (June).* Madrid.

——. 2019. *International Tourism Highlights – 2019.* Madrid.

——. 2019. *UNWTO Tourism Highlights.* Madrid.

——. 2019. *Yearbook of Tourism Statistics dataset [Electronic].* Madrid (data updated on 11 July 2019).

——. 2018. *Tourism and Culture Synergies,* UNWTO, Madrid.

——. *COVID-19: Measures to Support the Travel and Tourism Sector.* https://www.unwto.org/covid-19-measures -to-support-travel-tourism.

——. *COVID-19 Related Travel Restrictions Report – A Global Review for Tourism (Eighth report as of 2 December 2020).* https://www.unwto.org/news/70-of-destinations-have-lifted-travel-restrictions-but-global-gap -emerging.

——. *Tourism and Coronavirus Disease.* https://www.unwto.org/tourism-covid-19-coronavirus.

UNWTO and Government of Bangladesh. *Conference Report: Developing Sustainable and Inclusive Buddhist Heritage and Pilgrimage Circuit in South Asia's Buddhist Heartland.* 27–28 October 2015. Madrid.

UNWTO and Secretariat of State for Tourism of France. 2001. *Thesaurus on Tourism and Leisure Activities.* Madrid.

UNWTO and World Travel & Tourism Council. 2014. *The Impact of visa facilitation in ASEAN member states.* Madrid.

UNWTO Executive Council. *Thematic discussion: The Role of Tourism Routes in Fostering Regional Development and Integration (Session 98, 2 May 2014).* Santiago de Compostela.

World Bank. 2012. Case Study 1: ADB South Asia Subregional Economic Cooperation Tourism Development Project. http://documents1.worldbank.org/curated/en/128481563447559559/pdf/ADB-South-Asia -Subregional-Economic-Cooperation-Tourism-Development-Project-Case-Study-1.pdf

World Bank. 2020. World Development Indicators. http://datatopics.worldbank.org/world-development-indicators/ (accessed 15 June 2020).

World Economic Forum. 2019. *The Travel & Tourism Competitiveness Report 2019 - Travel and Tourism at a Tipping Point.* http://www3.weforum.org/docs/WEF_TTCR_2019.pdf.

——. 2019. *Travel & Tourism Competitiveness Index 2019.* Davos.

——. 2017. *Incredible India 2.0 India's $20 Billion Tourism Opportunity.* Developed by Bain & Company. September. Delhi.

——. 2017. *Travel & Tourism Competitiveness Index 2017.* Davos.

——. 2015. *Travel & Tourism Competitiveness Index 2015.* Davos.

World Food Travel Association. 2020. *2020 State of the Food Travel Industry Report.* https://worldfoodtravel.org/download-state-of-food-travel-industry-report/.

World Health Organization. Rolling Updates on Coronavirus Disease (COVID-19) – updated 31 July 2020. https://www.who.int/emergencies/diseases/novel-coronavirus-2019/events-as-they-happen

World Travel & Tourism Council (WTTC). 2020. WTTC Data Tool. https://tool.wttc.org/. (accessed 15 June 2020).

——. 2019. *Economic Impact of Travel and Tourism.*

——. *WTTC Economic Impact Report,* 2019. London.

——. 'Safe Travels': Global Protocols & Stamp for the New Normal. https://wttc.org/COVID-19/Safe-Travels-Global-Protocols-Stamp.